D0906552

Successful Project Management

Successful Project Management

A Step-by-Step Approach with Practical Examples

Third Edition

Milton D. Rosenau, Jr.

John Wiley & Sons, Inc.

NEW YORK / CHICHESTER / WEINHEIM / BRISBANE / SINGAPORE / TORONTO

This publication is designed to provide accurate and authoritative information in regard to the subject matter covered. It is sold with the understanding that the publisher is not engaged in rendering professional services. If professional advice or other expert assistance is required, the services of a competent professional person should be sought.

Library of Congress Cataloging-in-Publication Data
Rosenau, Milton D., 1931-
    Successful project management : a step-by-step approach with
practical examples / Milton D. Rosenau, Jr.—3rd ed.
        p.   cm.
    Includes bibliographical references and index.
    ISBN 0-471-29304-0 (cloth : alk. paper)
    1. Industrial project management.   I. Title.
HD69.P75R67   1998
658.4'04—dc21   98-17751

Printed in the United States of America.

10 9 8 7 6 5 4 3 2

Contents

Preface

WHY PROJECT MANAGEMENT IS IMPORTANT

Today much of the nonrepetitive work in corporations and other organizations is carried out in projects. Such project work is often a crucial ingredient in the strategic management of the enterprise. Whether or not they are central to strategy, projects are ubiquitous. The goal may be to rapidly commercialize a new product or service, install new capital equipment in your factory, develop a new software capability, redesign a process or procedure, benchmark some business practice, rearrange a facility, build something for another organization, or satisfactorily complete some other time-limited undertaking. In short, a project is an organized effort to accomplish a goal. The projects may be self-funded by your organization or performed under contract to another organization. Several organizations may be simultaneously involved. They may be small or large efforts and of short or long duration. Since 1981, when the first edition of this book was published, project management has experienced great professional growth, and there is a recognized body of knowledge, in part because more work is carried out in some form of project.

Improved management information systems have eliminated many of the middle management roles, allowing organizations to reduce or eliminate the ranks of middle management. Concomitantly, the era of purely functional organizations in which managerial command and control dispensed assignments and generally dictated how these were to be carried out is largely over. Now you are most likely to be working as part of a multifunctional team trying to accomplish some team goal, that is, seeing a project through to successful completion. You and your colleagues on the team may even be *personally accountable* for achieving the result in a timely manner!

WHO THIS BOOK IS FOR

This book is for you if you have been trained in a professional skill and now find yourself thrust into a position of managing a project in which you've had little or no direct experience. Now you must not only demonstrate your professional skill, but must take responsibility for schedule and budget, using physical and human resources over which you may have little or no real control. You may be an engineer (or scientist) asked to take responsibility for an engineering project, a product manager responsible for a new product development effort, a computer programmer assigned lead responsibility on a software engineering project, a civil engineer obliged to manage a construction project, or an office administrator charged with coordinating a facility relocation. No matter what your previous skill, if you're now responsible for getting something done by a specified date and with a limited budget, this book will help you grasp and master key practical skills for successful project management.

Although this book is aimed primarily at the newly appointed manager of a project, other professionals considering such a position will also find much of value. Project managers with limited project management experience will gain a better understanding or learn new techniques to improve their effectiveness. Functional managers who must support project managers will gain important insights.

Most of the discussion and examples are specific to small and medium-sized projects, rather than to very large projects (such as a space station or a major new sports arena). However, large projects are normally managed as a collection of smaller, integrated projects; so the book has relevance regardless of project size.

THIS BOOK'S APPROACH TO PROJECT MANAGEMENT

No treatment can or should make trivial a subject as complex as project management, but this book is specifically designed to make it as simple as possible. I provide a step-by-step approach that I have found effective in more than forty years of industrial experience with new product development for consumer and industrial markets, chemical formulation, engineering, government contract, research, management consulting, and volunteer organization projects. Since founding my management consulting practice in 1978, I have also had the benefit of teaching thousands of executive education and management seminar participants, working adults of all ages from a wide variety of industries. By taking you through the project management process in twenty-two steps (Chapters 2-23), the book aims to equip you with detailed tools that you can immediately apply to your first (or next) project and that will help you overcome the pitfalls that typically bedevil the project manager. These tools will help you deal with *both* technical or mechanical situations (such as the work breakdown structure, the schedule, and the budget) and conceptual or people issues (such as the concerns of top management and motivation).

USEFUL AND UNIQUE FEATURES OF THIS BOOK

There are many books on single, specialized aspects of project management (for instance, PERT and CPM), the use of some particular project management software (for instance, *Microsoft Project for Dummies*), and several books on project management in specialized applications (for instance, the construction or aerospace industries), but this book is intended specifically for the beginner who manages projects of any description. The first and second editions proved the value of the book's approach to explaining project management and helping readers improve their own skills; this third edition clarifies some of the previous material, brings it up to date, and eliminates some material made obsolete or irrelevant by the growing power, convenience, and ubiquitous availability of computers and useful software. This edition also provides numerous illustrations of pertinent output from a particular computer project management software package (primarily Microsoft Project 4.1 for Windows 95, but including discussion of some of the new capabilities in Microsoft Project '98). Microsoft Project, which is now widely used, has become almost a de facto standard for many companies. Despite these illustrations and my encouragement to use a suitable project management software package, this book is not a teaching manual about how to use that software, and I continue to emphasize that using project management software is not the same thing as being a project manager.

Each short chapter is devoted to a single topic and can be absorbed in one to two hours. Thus, the entire book can be mastered in a single month, making it uniquely useful for the working adult.

The book covers the subject matter chronologically, from a project's beginning to its end, and includes an example project that is developed step by step.

The book will be useful for almost any type of project and has many practical examples. In addition, I use some very basic illustrations to assure that a few critical issues are easily grasped, regardless of the reader's specific technical specialization.

Typical problems that may be encountered are highlighted in each chapter.

There are three appendices, containing abbreviations, a glossary of terms, and helpful checklists for project managers.

In addition, the book contains many illustrations of computer project management and other software output to demonstrate how this can be used. Employing such software will not make you a successful project manager, and utilizing it is not the same as being a project manager. Nevertheless, you can be a more effective project manager if you use such software in situations where it will be helpful to you. The examples are intended to illustrate some ways to use this software to your advantage. In addition, Chapter 26 provides some general guidance and cautions. Although my illustrations come from a specific mass-market project software package (Microsoft Project), my goal is not to demonstrate this product per se or teach how it may be used. Rather, I want to explain a few of the key useful aspects of this class of software.

HOW THIS BOOK IS ORGANIZED

The step-by-step approach divides the management of projects into five general managerial activities (or processes) and emphasizes the importance of satisfying the three constraints of performance specifications, schedule, and budget. The chapter sequence is a good match for the chronological concerns during a typical project. Your situation may differ, in which case you may wish to read the chapters out of sequence. Also, you may not be responsible for acquiring projects—they may be delegated to you—in which case you may wish to skip Chapters 3 and 4 until your responsibilities change.

The first managerial activity is defining the project. Under this topic, I introduce the Triple Constraint and discuss starting successful projects and negotiating and writing contracts.

The second activity is planning. Chapters covering why and how to plan, the work breakdown structure, scheduling tools and network diagrams (PERT, CPM, precedence, and others), time estimating, planning the budget, the impact of limited resources, and contingency and risk are included in my discussion of planning.

Leading is the third managerial activity, and as part of this topic I discuss how to organize a project, organizing the project team and the support team, the role of the project manager, and some practical tips for project managers.

The fourth activity is monitoring. Monitoring tools, reviews of projects, project cost reports, handling changes, and solving the inevitable problems are covered under this subject.

Under completing, the fifth managerial activity, I tell you how to complete a project and accomplish the final wrap-up.

Milton D. Rosenau, Jr., CMC, FIMC

Certified Management Consultant
Rosenau Consulting Company
Bellaire (Houston), Texas

1

What Is a Project?

This chapter differentiates project management from other managerial activity by examining how projects differ from other activities. It describes the sequence of five managerial activities that characterize project management.

This book is about project management. Because there is a proliferation of computer-based project management software—mass-market and other—it is crucial for you to realize that using such software is not the same thing as managing a project. As I point out in this chapter, project management involves many considerations. Because project management software is a tool that can be helpful, I provide illustrations of its utility throughout the book. It is also a trap for the unwary, as I clarify (in Chapter 26). In short, use it; do not abuse it.

Managing a project involves much more than using project management software.

DISTINGUISHING CHARACTERISTICS OF PROJECTS

There are four characteristics of projects that, taken together, distinguish project management from other managerial activities. Projects have a three-dimensional objective, are unique, involve resources, and are accomplished within an organization. These characteristics are discussed in the following sections.

Objective

Projects have a three-dimensional objective, which is the simultaneous accomplishment of the performance specification, the time schedule, and the cost budget. I call this the Triple Constraint and describe it in more detail in Chapter 2. Successful project management requires that the Triple Constraint be measurable (this is, specific and verifiable) and attainable. In fact, it is crucial that the people who are going to carry out the project work know how to attain the Triple Constraint point.

Uniqueness

Each project is unique because it is carried out only once, is temporary, and (in almost every instance) involves a different group of people.

Even though a second software project to write an accounts payable system is very similar to a first such project, there will be some differences, perhaps something as simple as the format of reports. The same is true of digging two ditches (the purpose or terrain may vary) or organizing two conventions (the sites or programs may differ), and so on. Because they differ, there is always some degree of imperfect understanding of exactly what is involved and what it will take to be successful.

Also, because projects are temporary, there are always authorization uncertainties (for instance, when a project will start and the exact scope of work to be carried out). In addition, a project does not go on forever. It starts when the first person goes to work and ends when the last person's work is finished. Somewhere in between, several or many people may be involved. Recruiting people to work on a project is similar to staffing a company that ultimately intends to go out of business.

Finally, the people who work on one project are rarely those who have worked on a previous project. Thus, friendships or antipathies are being created with significant consequences. Strong friendships may make people reluctant to end a project because they will no longer be working together. Conversely, antipathies may make it extremely difficult for two people to cooperate during the project.

Resources

Projects are accomplished by resources, namely, people and things. Many of the required resources are only marginally under the effective control of the project manager. For example, a required lathe may be controlled by a model shop group or a required computer may be controlled by a data-processing group.

The project manager must organize the correct human resources to take advantage of the available physical resources. Then the project manager has to deal with the constraints and emotional problems inherent in their use while trying to accomplish the project initiator's technical performance goals within the schedule and budget.

Managing projects means managing people. Managing people is often the most difficult aspect of managing a project, especially for recently appointed managers whose academic training is primarily in a technical discipline such as engineering, computer science, or even construction management. Such people tend to be more comfortable with things and numbers than with people.

Thus, you have to avoid the technical expert's propensity to concentrate on the quantitative aspects (for instance, engineering analyses or task budgets)—although these are not unimportant—and instead become more oriented to making things

PLAN

ACTUAL

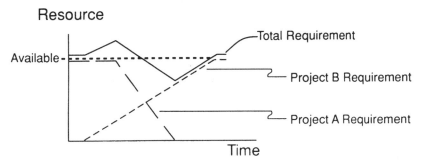

FIGURE 1-1. *Resource conflicts are inherent because the actual project resource requirements never produce a workload (on people in general, people of a particular skill, or facility) that perfectly matches resource availability.*

happen through people. Many good technical experts make poor project managers because they can't deal with the intangibility of people issues, such as the need to "sell" (and resell) a project to other managers. Some project managers recognize this need but cannot communicate effectively.

Resource conflicts are inherent, as illustrated in Figure 1-1. In this simple illustration, there are only two projects illustrated. Their need for a particular resource (square feet, hours of usage of a DNA synthesizer, personnel in general, or a specific kind of human resource, for example, junior level analytic chemists) is shown. A planned requirement exactly matches the available resource. However, the timing of one project is slightly altered, and as a result, the required resource no longer matches what is available. First there is too much work, and later there is not enough work. Resource overloads and underloads are a common—and bedeviling—characteristic of the project manager's world.

These mismatches are bad enough where the resources are physical things. The resulting problems are often much worse where the resource mismatch (be it overload or underload) is people.

<hr>

The project manager must spend lots of time with people.

<hr>

Because of the tremendous importance of these people issues, I thoroughly discuss organizational options and their influence on resource control and availability in Chapter 12. The rest of Part 3 is devoted to an extensive review of effective actions you can take as project manager to lead both the project and support teams.

Organization

Every organization has a multiplicity of purposes at any given moment, if for no other reason than that it is composed of many individuals with varied skills, interests, personalities, and unpredictabilities. Thus, the project manager will often be frustrated by the many other directions in which the organization seems to be (and often is) moving (see Figure 1-2). These multiple directions arise because of personal aspirations and interests, because of various parochial interests by different components of the organization, and because of many projects being carried out simultaneously.

As Figure 1-2 suggests, customers and users (who may be your customers, customers of your customers, or other people) will also affect your organization. Thus, in many instances, the project manager (even with a loyal and dedicated team) may be frustrated in trying to achieve the project goal.

Project management, in large part, is the management of interpersonal conflict, which is inherent in complex organizational situations.

OTHER ASPECTS OF PROJECTS

There are four other aspects of a project that may alter the difficulty: origin, what is produced, the marketplace, and the size.

Origin

Imagine you are a mechanical engineer and your boss has asked you to manage a materials study project. Or imagine you are a computer programmer who has been asked to set up a new financial report, a civil engineer asked to manage a construction project, or a facilities administrator asked to manage the relocation of a division into a new building. How would you go about it?

Before reading further, consider this hypothetical situation. How does this activity differ from others in which you might engage? Is it similar to going to your workplace each day? In what ways does it differ?

At this point, you probably want more information about the project. For instance, is it small or large? What do you mean by small or large—the quantity of materials

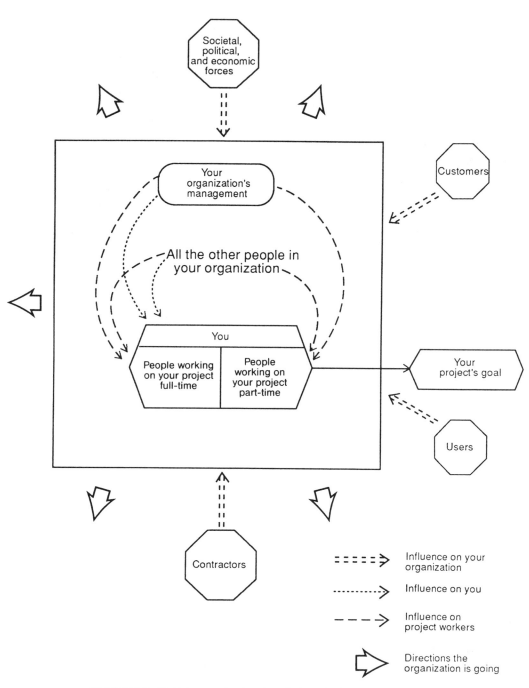

FIGURE 1-2. *Multiple organizational goals impede the project manager.*

to be studied, the deadline, or the budget? How does project magnitude alter your approach?

This hypothetical project illustrates some of the characteristics that distinguish projects from other activities. Projects originate because something not done before must be done. Although going to work on some mornings may seem to be a major undertaking, it is not usually considered a project. Going to work is an activity that repeats a prior activity, namely, going to work the day before. This aspect of the

---

Projects are one-of-a-kind undertakings that originate when something has to be done.

---

definition of a project is not clear-cut. If your project had been to build an amplifier circuit, at some point, building a second or third or fourth amplifier circuit ceases to be a project and becomes a repetitive activity (for example, electronic assembly). As a general rule, if the amplifiers are different from one another, then each is a project. If each amplifier is virtually identical, we have a production line for amplifiers and are not engaged in a project per se.

Product

There are hardware and software projects.

There are many ways to characterize projects. "Hardware project" and "software project" are common terms, depending on whether the final result is a tangible product (hardware) or a report or some other form of documentation (software). There are important distinctions within these broad categories. For instance, capital construction projects include nuclear power plants (with significant political, regulatory, and safety considerations), process plants (with differing issues, depending on whether they are to handle liquids or solids), and other projects (commercial buildings, power plants, civil works, manufacturing facilities, and so on, all of which have special considerations).

Thus, the product or end result of a project is a second characteristic. A project is not an ongoing activity but rather an undertaking that ends with a specified accomplishment. In the first hypothetical example, this would be the completion of the materials study project.

Marketplace

Projects can be categorized by their source of sponsorship, as shown in Table 1-1. One of the fringe benefits of being a good project manager is that the tools and techniques applied on the job also have utility for personal undertakings. Figure 1-3 is a time-based activity-on-arrow network (discussed in more detail in Chapter 8) that my wife and I used when we moved from one house to another. Because we only had a thirty-day escrow period, scheduling of many activities in the correct sequences was critical, and we benefited greatly from use of this particular tool, even though this was a family project.

If the customer is outside your organization, the project may be undertaken as a contract, and satisfying the customer's needs will be critical. If the customer is a

TABLE 1-1   Project Types

Source	Project
Personal or family	Dig a ditch for roses
	Repair garage door opener
	Add a new utility room
Company sponsored—for the company itself	Set up a new computer
	Put a new plant into operation
	Commercialize a new product
Customer sponsored	Construct new shopping center
	Design and build mach 3 supersonic transport
	Produce pressure-sensitive postage stamps for the postal service
Government sponsored—done by a governmental entity	Issue new $1 coin
	Produce the surgeon general's report
	Redesign income tax forms

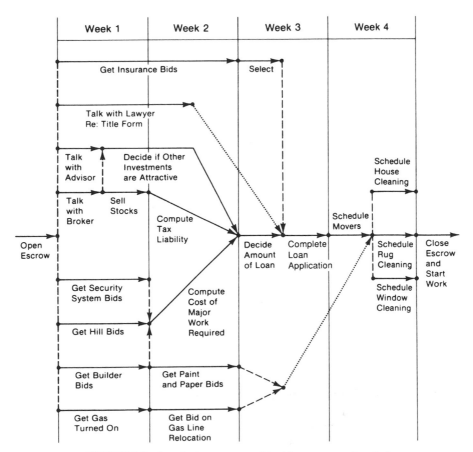

FIGURE 1-3. *A project management tool for a personal project.*

governmental entity, there will probably be very formal procedures. If the project is done for and inside your organization, no formal legal contract will be required, but the timeliness of completion and the expected return on investment are frequently critical. If your company wishes to conduct a study of new materials for its own information, the process may be very informal. Regardless of whether your customer is inside or outside your organization, it is wise to act as if the customer is external and to define the objective both formally and clearly.

For whom a project is done affects how it is done.

The nature of the competition is also important. The only materials laboratory in a region is in a favorable situation. A materials laboratory that is one of many in some small city is in a very different and less favorable competitive situation.

Once an organization or a project manager has successfully concluded a project, a company-sponsored software project, for instance, it will be much easier to undertake a second project of that sort. Although there are similarities in managing all kinds of projects, successfully completing a company-sponsored software project is very different from successfully completing a customer-sponsored hardware project.

In general, projects are done by someone or some organization for someone else or for some other organization. I prefer to call the requester the project sponsor. The sponsor may be your boss, someone else (perhaps a peer) within your organization, or a customer (a single person or organization).

Size

"Program" is commonly used synonymously with "project." Thus, the expression "program management" is often used interchangeably with "project management." Some organizations use "task management" as well. Program management, project management, and task management are generally identical. But programs are usually larger than projects, and projects are usually larger than tasks. Thus, there is some connotation of size when terms other than project management are used. Nevertheless, the techniques and methodology are essentially the same, differing only in detail. I shall use "project" throughout the rest of this book.

The trans-Alaska pipeline and the manned lunar landing projects required many years and billions of dollars. The space shuttle project is also both long and very costly. Conversely, a computer programming project may be completed in a few hours or days. Hence, size or complexity does not distinguish a project from another activity. They do, however, affect project success.

THE PROJECT MANAGEMENT PROCESS

Project management requires five different managerial activities and can thus be most simply structured as a five-step process:

1. Defining—defining the project's goals
2. Planning—planning how you and your team will satisfy the Triple Constraint (goal) of performance specification, time schedule, and money budget (The plan depends on the mix of human and physical resources to be used.)
3. Leading—providing managerial guidance to human resources, subordinates, and others (including subcontractors) that will result in their doing effective, timely work
4. Monitoring—measuring the project work to find out how progress differs from plan in time to initiate corrective action (This often leads to replanning, which may force a goal [definition] change, with a consequent need to change resources.)
5. Completing—making sure that the job that is finally done conforms to the current definition of what was to be done and wrapping up all the loose ends, such as documentation

The first two steps are not necessarily separate and sequential, except when the project initiator issues a firm, complete, and unambiguous statement of the desired project output, in which case the organization that will carry out the project may start to plan how to achieve it. It is more common to start with a proposed work definition, which is then jointly renegotiated after preliminary planning elucidates some consequences of the initially proposed work definition. The definition must be measurable (specific, tangible, and verifiable) and attainable (in the opinion of the people who will do the work) if you want to be successful. Being successful also requires that management agrees the project is justified and that the resources the project team needs will be available. Consider the predicament of one project manager.

> **Both the project team and appropriate management must agree that the project's objective is attainable.**

> I am a project manager working with four or five other people. Our project consists of redesigning a piece of hardware, making any improvement possible, and then actually making five to ten prototype systems. Our customer has not specified any time limit or deadline for our project to be completed. All that is expected is to accomplish "something" every calendar quarter. I know the day will come when the customer will say that he wants it now!

> My problem is that I don't seem to be able to motivate my people to work at a little faster pace. I have tried setting goals with each of them individually; I have tried letting them set their own goals; I have tried to set deadline dates for a particular part of the job to be done; but nothing seems to work. I have tried to work with them and actually help do their work. They seem to get their part of the job done when they get around to it.

Thus, in fact, the resources to be dealt with in the leading phase often must be

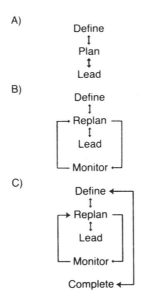

FIGURE 1-4. *The five activities are interdependent.*

considered before planning can be finished (see Figure 1-4A). For instance, you might need engineers familiar with carbon fibers if the plan for a materials study project includes the study of that kind of material, whereas you would use a metallurgist if the project were to study only metals.

No project goes in accordance with your plan. What you don't know when you start is where it will go awry. Consequently, as you will see in later chapters, replanning is almost always required, thus frequently amending the negotiated definition (see Figure 1-4B). Ultimately, the project can be completed when the work that is done satisfies the current requirement (see Figure 1-4C).

Projects require five steps, which may overlap.

Nevertheless, the five-step managerial activity process covers each required action and is a useful conceptual sequence in which to consider all project management. Thus, this book is organized according to it.

The Triple Constraint, an extremely important notion for project management, provides the defining parameters of a project. It consists of three dimensions:

1. Performance specification
2. Time schedule
3. Money budget

(In some projects, most commonly research and development or computer software projects, the budget may only be stated in terms of labor hours, rather than dollars. Nevertheless, labor hours are equivalent to dollars when hourly labor rates are applied.) All projects are defined and characterized by a Triple Constraint.

TYPICAL PROBLEMS

Failure to identify a project for what it truly is usually leads to missed specifications, late completion, and/or a budget overrun. The solution is to recognize that there is a project when something must be done and then to organize to complete the project in the least disruptive way.

The fact that humans are involved in projects and must be worked with is often especially troubling for technically trained project managers. In many cases, this need to work with people is the biggest obstacle that a good technologist encounters. Such technically trained managers expect, but do not get, completely logical or rational behavior from these people (or, for that matter, from themselves). Finally, although many project management tools (for instance, many of the planning and monitoring tools discussed in this book) are completely rational, project management in an overall sense is not an exact science.

Successful project management means meeting the performance specification (that is, objective or technical goals), on schedule, and within the budget.

MATERIALS STUDY PROJECT

I shall use the project introduced here throughout the book to illustrate some of the issues discussed in each chapter. Imagine you are a mechanical engineer, Mel Chase, working for the Successful Project Management Company (SUPROMAC). Your boss, Mike Miller, has asked you to act as project manager on a materials study project a customer (NERESCO or Needs Results Corporation) wishes to have SUPROMAC perform. Because this is a one-time effort, not identical to anything SUPROMAC has done before, project management is clearly going to be applicable.

HIGHLIGHTS

- Projects are temporary undertakings with a specific objective that are accomplished by organized application of appropriate resources.
- Four distinguishing characteristics of projects are a three-dimensional objective, uniqueness, the involvement of resources, and interaction with the organization.
- Other aspects of projects are their origin, product, marketplace, and size.

- Size and complexity do not distinguish projects from other activities.
- Project management is the process of achieving project objectives in any organizational framework despite countervailing pressures. This often requires selling and reselling others on the project's importance.
- There are five managerial activities in project management: defining, planning, leading, monitoring, and completing.
- The Triple Constraint defines a project: performance specification, time schedule, and money budget.

Part 1

Defining the Goals
of a Project

2

The Triple Constraint

This chapter introduces the concept of the Triple Constraint as a project definition, identifies some of the obstacles to satisfying it, and describes some steps to help achieve it. The consequences of various project outcomes are considered from the point of view of satisfying the Triple Constraint.

THE CONCEPT OF THE TRIPLE CONSTRAINT

Figure 2-1 illustrates the Triple Constraint, a very important concept I shall emphasize throughout the book. Successful project management means accomplishing the performance specifications on or before the time limit and within the budgeted cost. The cost is usually measured in dollars, francs, marks, or whatever the coin of the realm but may sometimes be measured in the number of labor hours, or perhaps labor hours in each of several labor rate categories such as senior research chemist and laboratory technician; carpenter, crane operator, and stonemason; system analyst and programmer; senior engineer, engineer, and junior engineer; and so forth. The key point the Triple Constraint illustrates is the need to simultaneously satisfy three independent goals—not just one.

> **The Triple Constraint defines projects: performance specifications, time schedule, and money (or labor hour) budget.**

Money may actually appear in two dimensions of the Triple Constraint. It is obvious that it appears in the cost dimension, where it may include not only actual project expenses but, in some situations, capital expenses (even though this may be in the organization's capital budget as well). Normally, the Triple Constraint's cost dimension involves the planned or actual project expense, usually including the overhead and the general and administrative (G&A) burdens. Money may also appear in the performance dimension, where it might be stated as the factory (that is, standard) cost of a manufactured product, the normal operating cost of whatever

15

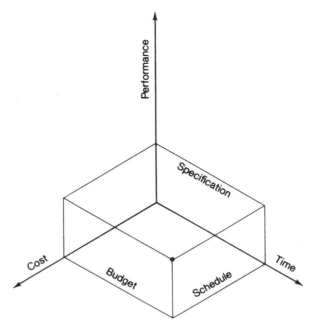

FIGURE 2-1. *The Triple Constraint.*

the project produces, or the required maintenance cost of whatever the project produces. Earned value (EV), which may also be stated in dollars, is a progress measure that your project may require; this is discussed in Chapter 19.

Figure 2-2 illustrates the practical relationship between the three Triple Constraint parameters. At a given level of performance, a particular schedule will result for a particular budget. If more budget is available (which would be represented by a lower curve than is drawn in the figure), more effective resources may be applied to the project, and the schedule can be shortened. The figure also illustrates that at some high level of performance, no one currently knows how to accomplish the very high performance objective, (that is, the budget curve becomes exceedingly steep, and the resulting schedule becomes infinite).

The schedule can be shortened if a higher budget allows more effective resources to be used.

OBSTACLES TO SATISFYING THE TRIPLE CONSTRAINT

Unfortunately, the Triple Constraint is very difficult to satisfy because most of what occurs during any project conspires to pull the performance below specification and to delay the project so it falls behind schedule, which usually makes it exceed budget. Because no project follows the plan, the successful project manager must be alert to potential problems to achieve the Triple Constraint.

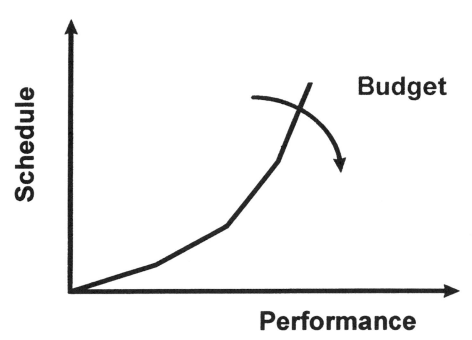

FIGURE 2-2. *The Triple Constraint trade-off.*

Under the best of circumstances, it is difficult to achieve the Triple Constraint. However, it is also normal for the Triple Constraint to change during the course of the project. For example, the sponsor (who may be a customer, your boss, or a user) may demand or request a changed objective. In some instances, you or the project team may become convinced the original Triple Constraint cannot be achieved and will have to propose an alternative end point. Finally, a change may be imposed by, for instance, new legislation or regulation.

As you can see, projects encounter a wide variety of problems. Some of the principal ones are enumerated in the following sections, organized by the dimension of the Triple Constraint most affected. Sometimes only general symptoms are evident, rather than specific problems with a single dimension of the Triple Constraint. Although Figure 2-1 shows these three dimensions to be orthogonal axes, hence mutually exclusive, project management is not that ideal. For instance, the materials study project result may appear poor compared to expectation in relation to the money invested. Similarly, a late project, for example, will usually also overrun the budget. Thus, items listed as causes of time problems may also cause cost problems.

Performance Problems

There are many reasons the performance specification is difficult to achieve. I shall discuss three of them. First, there may have been poor communication between the contractor and the customer. That is, they have different perceptions of the specification, or the wording is ambiguous. For instance, "security" means different things to different people or in a different context. To a military contracting officer, it may mean a secret classification, whereas a person working on a military software project might interpret it as meaning data protection. At the Social Security Administration, the word may mean enough money to live well, whereas in the brokerage industry, it might mean stocks or bonds. In short, capable, dedicated, and perfectly honest people may interpret a simple word differently—with attendant problems. This is not an academic difficulty. In early 1990, a $150 million communication satellite was stranded in a useless orbit due to a wiring mistake that resulted from two groups interpreting language differently.

Today, quality is generally viewed as satisfying users' needs by correctly defining their requirements. I consider quality—however defined—as being a part of the performance dimension. Poor or ambiguous communications between the sponsor (or customer) and project team (or contractor) may lead to a quality disappointment. This quality shortfall can occur even if other performance specifications are well satisfied. For example, a "good quality finish" on a product's surface may not be precisely measurable, and the result may be either a disappointment or unacceptable.

Clarify unclear specifications.

A second problem arises because the contractor's or the customer's assumptions have been too optimistic. Their goals may have been too ambitious, which is not uncommon in advanced technology.

Third, the contractor may do a poor design job or make mistakes in executing contract performance. Unfortunately, workers (and managers) make errors occasionally, and these errors may cause a performance deficiency.

Time Problems

Schedule problems arise for several reasons, the most insidious being an overemphasis on the performance dimension at the expense of a balanced view of the Triple Constraint. For instance, scientists or engineers, who are commonly appointed project managers, tend to concentrate on the technology and to strive for technical innovations or breakthroughs. A computer programmer may emphasize work on a clever algorithm or use a new programming language rather than expeditiously completing the program with existing capabilities. Such striving is accomplished at the expense of the schedule, and it frequently has unfavorable cost repercussions. To put this another way, "better" is the enemy of "good enough."

Technical excellence usually interferes with meeting the schedule.

Even where a fascination with technology is not overwhelming, technically trained people, such as engineers, tend to assume the performance specification is sacrosanct, whereas they consider it permissible to miss the schedule or budget objectives. Conversely, a customer might be satisfied (if not ecstatic) to achieve 90 percent of the performance specifications, provided both the schedule and the budget are met (for example, on a planetary exploration project, it is vital to be ready by

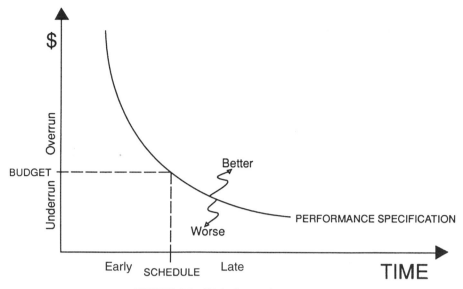

FIGURE 2-3. *Triple Constraint outcomes.*

the "launch window," a restricted range of dates, even if one of several nonessential instruments must be omitted).

Figure 2-3 illustrates various Triple Constraint outcomes and their relationships. For example, performance that is better than the specification is normally obtained only if there is a budget overrun or a late delivery, and most frequently when both occur. This illustrates why it is so important for the project manager to moderate the enthusiasm of technical experts, who can always see the possibility of doing better and are eager to achieve higher performance.

A second source of difficulty in meeting the schedule arises because resources are not available when required. These resources may be either equipment (such as lathes or computer hours) or personnel (a well-qualified circuit designer, for example). This absence of planned resources forces the project manager to find substitutes, which may require a subcontract to get some design work done. Or it may mean using marginally qualified people who take longer to complete a circuit design than the well-qualified person previously assumed to be available.

Third, a project can get into schedule difficulty because those assigned to it are not interested in their tasks. In this case, they may choose to work on other things or work half-heartedly on the project.

Fourth, schedule problems can arise because the performance specification is raised. For instance, increased efforts that lead to additional work are accepted. A common occurrence that illustrates this is a customer asking for a few extra control switches. The project manager may misperceive this as trivial because a control panel is already being provided. If he or she agrees to provide these extra switches, which

The time schedule should change if the performance specifications change.

were not part of the original proposal, the project manager is agreeing to do additional work (perhaps even modifying the control logic) without changing the schedule (or cost). There is, however, additional work called for to put in the switches, and it does not take many changes of this sort first to produce a one-day schedule slippage and then a one-week and so on until the project is in serious schedule difficulty.

Cost Problems

Cost problems arise for many reasons. When a project is in trouble on its time dimension, it will often be in trouble on its cost dimension as well because resources are not being used as efficiently as planned.

A second cause is the "liars contest" that often occurs during contract negotiation (especially if the project is done for another organization). Imagine you have bid $10,000,000 to build a shopping center. During negotiations you were told that unless your price is lowered to $9,500,000, the contract will be awarded to another company. In your desire to obtain the work, you and your management agree to minor wording changes, which appear to reduce the scope of work a bit and permit you to justify a substantial cost reduction. When you make this kind of price reduction without fundamental work reductions, you have built in a cost overrun at the very beginning of the project. A professional, experienced project manager will never agree to this kind of negotiation unless he or she knows the money will be restored later in contract changes. This is a potential cost problem for which the less experienced project manager must be very alert.

> **Do not "buy in" without a "get well" plan or a performance reduction.**

This liars contest also happens inside an organization. In this case, the project has to be sold to upper management, and you are competing with other organization managers for authorization. Chapters 3 and 4 are primarily about doing work under contract for another organization, but they cover much that is important for managers of projects being done inside their own organizations. As I said in Chapter 1, it is smart to pretend your customer and the users are from another organization—and to define the project with all the formality you would use for such an external contract.

A third source of cost difficulties arises because many of the initial cost estimates are simply too optimistic. They do not reflect the inefficiencies that will occur in scheduling resources to perform the work or the fact that less well qualified people may be assigned to do the work.

Occasionally, mistakes are made in the cost estimating. Like design mistakes, these are unfortunate, and careful scrutiny and review can minimize this occurrence.

A fifth reason for cost problems is simply an inadequate cost consciousness on the part of the project management or a failure to have an adequate cost management system. This is never excusable.

> **Funding is required per plan.**

Sixth, funding may not occur according to plan, and this may produce problems. Suppose your project is to run for three months and the customer originally proposes to fund you with $100,000 per month (that is, provide progress payments). The first month has gone according to plan and you have received your $100,000. At the start of the second month, your customer informs you they have only $50,000 available for that month but they will have $150,000 available the third month. You will still get $300,000, but you cannot apply the planned resources during the second month and must rush to catch up during the final month. Obviously, this would be a less efficient way to operate and means your costs will be higher than originally planned. When confronted with this sort of refunding proposal by a customer, therefore, the prudent project manager will insist on a budget increase if the performance specifications and schedule are not to be changed.

Sometimes when you start a project, you and the customer agree that it is impossible to define all the specifications completely. Whenever this happens, it is crucial to agree specifically and in detail on what will be done by a given date. For instance, you may agree to work for one year to establish the complete, final specifications for the project. Then the second phase work can be the rest of the project, which is now aimed at specific, achievable goals. This approach avoids open-ended (thus endless, and hence overrun) projects. The goal you should strive for is a complete, clear, and specific end point.

PROJECT OUTCOMES

A project may end anywhere in the three-dimensional space illustrated by the Triple Constraint. Whether a deviation from the Triple Constraint point is acceptable depends on the project. For example, a project might be late. If the project is to design, test, obtain certification, and then install a vapor recovery system by a legislated date, lateness would be inappropriate. Or if the project is to construct the grandstands for the presidential inauguration on January 20, lateness would be an intolerable outcome, regardless of grandstand quality or cost. As suggested earlier, schedule compliance is also the overriding concern for planetary launch.

In other situations, the budget may be the most important issue. When a contractor accepts a fixed price contract to deliver a working piece of hardware, he or she will lose money if the budget is overrun. Conversely, on a cost reimbursable contract, the customer must pay for the overrun. In either situation, ability to pay is crucial.

Third, there are situations in which the performance specification may not be missed. Once again, the project specifics will determine whether this is crucial. In the early 1960s, before spacecraft could carry large instruments, the Stratoscope II project was undertaken. The goal was to carry a large telescope by balloon to eighty thousand feet, where it would be above almost all the earth's atmosphere, permitting the optical resolution of fine detail in distant nebulae. To accomplish this goal, the telescope's primary mirror had to be virtually perfect. Because the project could succeed only with such a mirror, there was no point in flying the balloon-borne telescope until this performance specification had been met. This required both more time than the original schedule and more money than the original budget. In other cases, there may be some law that must be complied with.

TYPICAL PROBLEMS

**Project specifics
determine the
relative importance
of each dimension
of the Triple
Constraint.**

The project sponsor's emphasis is almost always unclear initially, and project personnel tend to assume their own biases in ranking the relative importance of each dimension. This can easily lead to a disastrous outcome, which can be avoided by adequate discussions between the customer and the contractor at the project's inception.

A second problem is that many project managers are asked—or compelled—to take on responsibility for a project that was planned by someone else. In some such cases, they do not know how to accomplish the Triple Constraint. In such a situation, the new project manager must take some time to study the project's scope, offer several options to his or her management (and, perhaps, the sponsor), and provide a recommendation on how to proceed.

A third major problem, mentioned earlier, is the myopic attention to the performance dimension by technical personnel. It can be overcome, or at least reduced, if the project manager clearly conveys the customer's emphasis and its rationale.

MATERIALS STUDY PROJECT

In discussion with SUPROMAC about the NERESCO materials study project, Mel Chase realizes there are easily half a dozen materials that might be relevant to NERESCO's needs and that SUPROMAC might therefore reasonably examine. He feels this will require about one year and cost $1 million.

Discussions with NERESCO reveal that they require the study results in six months and their budget cannot exceed $500,000. Thus, some of the possibilities for Mel Chase to consider are the following:

Try to convince NERESCO that he is right and more materials should be studied, taking longer and costing more than they want.

Refuse to undertake the project.

Divide the project into several phases. The early phases can be very quick and low cost, requiring joint SUPROMAC-NERESCO review upon conclusion to set goals for later phases.

Explain the risk to NERESCO in studying only a few materials and jointly work with them to choose the specific (fewer) materials to be studied within their time and budget constraints.

One of the last two possibilities is probably preferable in this case.

HIGHLIGHTS

- The Triple Constraint defines all projects.
- The Triple Constraint consists of performance specifications, a time schedule, and a money or labor hour budget.
- Obstacles that prevent satisfying the Triple Constraint are not mutually exclusive.
- Project specifics determine the relative importance of each dimension of the Triple Constraint.
- Adequate and clear discussions among the customer, the project manager, and the technical personnel can help avoid many common problems.

3

How to Start Successful Projects

Proposals bridge the defining and planning activities on projects. This chapter examines the strategic issues that govern writing successful proposals. Although this chapter focuses on proposals written for an external organization, all the same principles apply to proposing to undertake projects within your own organization. Project proposals constitute the definition embodied in the contract or work authorization. A good proposal includes a thorough plan for work performance that embodies the Triple Constraint. This chapter also discusses the proposal process in detail.

STRATEGIC ISSUES

Some organizations lack strategic discipline. Any project is good or desirable in this environment. But even very large organizations have finite resources, and the more successful companies concentrate these on the few most attractive projects. To do otherwise starves some (or all) key efforts for critical capabilities. The goal of this section is to provide you with some orientation to a few of the concerns that confront senior managers in considering which efforts to pursue or authorize.

Framework

Figure 3-1 illustrates the strategic framework for obtaining winning projects. It does not matter whether the projects originate outside the organization by a customer or within the organization. However, many organizations dissipate their energies in preparing losing proposals, which

1. Cannot win the competition
2. Will have an unfavorable outcome with regard to the Triple Constraint

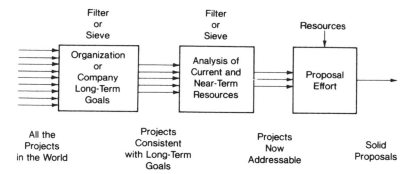

FIGURE 3-1. *Strategy to obtain winning projects.*

This material originally appeared in a videotape presentation produced by the Association for Media-based Continuing Education for Engineers, Atlanta, Georgia.

3. Can be successful but are insignificant or irrelevant to the proposing organization

Even with the filtering illustrated in Figure 3-1, you never know which specific proposal you will win (if they are written for an outside sponsor) or which will be approved (if they are for a sponsor within your organization). Thus, because you can never be certain which project proposals will be approved, the organization is always somewhat maladapted to the existing workload. This is one reason the workload does not match the available resources, as was illustrated in Figure 1-1.

Therefore, the basis of a successful strategy is to filter out losing projects. These include projects inconsistent with the organization's long-term goals or with the current and near-term resources within or otherwise available to the organization. Such filters might reject consumer project efforts in an industrial product company. Similarly, the filters could reject a fixed price contract for a technical development, scientific advance, invention, or other breakthrough in a conservative company that will not normally undertake a fixed price contract for something not previously accomplished. In one particular instance, a major aerospace contractor working on a $300 million contract expected final costs to be $600 million. Consequently, that contractor asserted they would never again take a firm fixed price contract for any development work from the Department of Defense.

Filtering out the huge number of possible projects the organization might have addressed leaves a much smaller number of projects that are appropriate for the organization to consider. It can then address some or perhaps all of these in proposal efforts to which it applies adequate and appropriate resources. The result of this process is that an organization submits only very much founded proposals for consideration. This result is most likely to be achieved if the organization has a careful review process, often called the "bid/no bid" decision.

Concentrate on meaningful prospective winners.

In summary, avoid projects that are inconsistent with your organization's long-term goals or current and near-term resources, unlikely to win the proposal competition, unlikely to satisfy the intended Triple Constraint, or insignificant or irrelevant. There is a special situation in which your immediate superior demands that you undertake a new effort but you and your department or group are overloaded. In this case, your response must indicate the nature of the problem, suggest some alternatives, and indicate how you plan to satisfy this new requirement.

Bid/No Bid Decision

The decision whether to bid on a proposal opportunity, whether to an external organization or within the organization itself, must be taken within the context of the organization's strategic framework. This framework is of course specific to the organization at that particular time. The company that rejects consumer product projects today may have the interest and capability to undertake these projects five years from now. There are many issues involved in this decision, four of which I discuss.

The Requirement

Clear criteria are vital.

The first issue is whether there is a real requirement. If a requirement exists, it is important to decide whether the funds are really available. It is possible for there to be a real requirement but no funds. For instance, many isolated rural villages in undeveloped nations could use an electric power generation system, but these villages and the nations of which they are a part cannot afford to pay for their installation. Contractors occasionally seek subcontractor proposals for "window dressing" to satisfy someone or otherwise justify an alternative approach they propose to take. Clearly, it is not worth responding to such requests for a proposal because the effort will not result in a winning project.

Proposals to external organizations may be written in response to a request for quotation (RFQ) or because you have identified an opportunity. Business you may choose to solicit from external sources is illustrated in Figure 3-2, and the process

Getting one project requires a lot of preproposal work.

is called the proposal funnel because you must keep injecting inputs at the top (that is, find more prospects) if you expect to get anything out of the bottom (that is, get more self-initiated projects). Again, not every possible project gets initiated, and there is no assurance that any amount of preproposal work will result in a specific project. W is a larger number than X, which is larger than Y, which is larger than Z, which is normally larger than one.

It is also important to examine the priority or importance of the proposed project. In this connection, the relationship of a particular proposal opportunity to present and future programs is also an important issue. Many small "paper study" projects, not attractive per se, actually are very attractive because they can lead in the future to large production programs.

FIGURE 3-2. *The "funnel" of effort leading to a solicited project. A measure of proposal efficiency is the size of the numbers Z, Y, X, and W.*

Project Value The project's social, ecological, and energy impacts might be highly significant to an organization in deciding whether or not to bid on a project. Presumably, a project to build the prototype of a device to clean up oil spills economically would be more attractive than a project that endangers the earth's ozone layer. Similarly, a project that offers the opportunity to apply important new technology or otherwise enhance the organization's reputation might have high nonmonetary value. Conversely, the expected sales of a commercial new product project effort and the profits of such an effort would also be significant issues.

Finally, many projects are merely a hidden obligation for an organization to accept future financial commitments for new capital or facility investments. This must be discovered before any money is spent so as to be certain that future major financial commitments are within the organization's resources and abilities as well as being consistent with the project's prospective value.

Response Ability The central issue here, also illustrated in Figure 3-1, is the organization's present capability first to prepare a winning proposal and second to perform the proposed work. If some capabilities are not actually present, there must be a viable plan to make them available when they are needed.

The organization must have the ability.

Winning the Competition First, an organization must ask whether there is advance information about the project available to it. This is particularly true about efforts arising from a customer organization and being presented to a contracting company as a request for a proposal (RFP), but it is also relevant for efforts within an organization. Lack of advance information often indicates that someone else has a head start or the request was hastily created and lacks substance.

A second issue concerns the customer. Is it an individual (yourself, a friend, or someone else) or a commercial organization? If the latter, is it your organization or an external organization? If it is your own company, has your superior ordered you to carry out the project or must your project proposal compete with other project proposals that management is considering for funding? If the customer is a governmental organization—a city, county, state, or federal (domestic or foreign) entity—there will probably be detailed specifications, formal quality standards, perhaps the necessity for surety bonds, and very often rigid and formalized inspection procedures. Who are the key personnel within the customer organization? Are they known to your organization? What history do you have with them or with their organization? Is your organization's reputation with the prospective customer favorable?

Your organization must have enough money to write the proposal and to sustain the postproposal selling and negotiating efforts. Therefore, you must know whether money to invest in this kind of activity is available. You must expect the project to earn more money than the proposal costs because you will not win every job on which you propose.

There are situations in which you will be the sole source recipient of an RFP, or, for instance, when your boss tells you to carry out a project within your organization. There is thus no competition, but how you perform the job is still important. It may be better to decline an effort when it is offered sole source if you are convinced the performance you can provide will be at best marginal.

All competition has to be analyzed. Some relevant issues are the competition's technical and managerial competence, its ability to produce the requested project output, an estimate of its interest in the particular type of project, its need (or degree of "hunger"), and its prior relationship with the customer or organization.

In summary, avoid projects that are inconsistent with your organization's long-term goals, inconsistent with your organization's current and near-term resources, unlikely to win the proposal competition, unlikely to satisfy the intended Triple Constraint, or insignificant or irrelevant.

THE PROPOSAL PROCESS

Although you will not win every proposal, you must be willing and able to satisfy any commitments you propose to undertake.

The proposal process entails more than writing the proposal (see Figure 3-3). That's the meat of the proposal sandwich. A sandwich consists of two pieces of bread surrounding the meat, and, analogously, the proposal also requires preparation work and postsubmission work. The goal of the preproposal work is to learn about the customer's problem and bias, which allows you to set a proposal theme. The proposal theme provides a focus to direct everyone contributing to the proposal, which increases the odds of producing a coherent, winning proposal. Thus, the overall proposal process includes the following:

1. Authorization (which formally considers the issues previously discussed)
2. Selection of a dominant theme
3. Preparation of the statement of work
4. Development of a plan to satisfy the Triple Constraint (an effort for which checklists may be helpful)

The Proposal Sandwich

The Finish
Postsubmission Efforts

The Meat
The Proposal
Document

The Foundation
Preproposal Work

FIGURE 3-3. The proposal sandwich.

This material originally appeared in a videotape presentation produced by the Association for Media-based Continuing Education for Engineers, Atlanta, Georgia.

5. Adjustment to remove inconsistencies and inadequacies
6. Approval
7. Submission
8. Postsubmission follow-up, including presentations and contract negotiations

Authorization

The proposal process frequently starts before an RFP, in which case the effort is frequently called the preproposal effort. Regardless of when the proposal process is initiated, this activity must be authorized. A form such as that in Figure 3-4 can be used for this purpose. In preparing a proposal, an organization is going to commit a certain amount of effort and money to it. This investment should be made only when it seems that the opportunity then available has a good chance of paying off and only if it is consistent with the organization's goals.

Another point to consider at the time of proposal authorization is the individual who will manage it. Ideally, the proposal manager should be the intended project manager. Often this ideal is not achieved. A proposal manager who knows that he or she will not be the project manager is likely to make commitments that are hard (or impossible) to meet, especially when the proposal manager's performance is judged on his or her ability to obtain (as opposed to manage) new business.

Consider, then begin. A proposal itself can be considered a project with a Triple Constraint, in which case the performance objective is submission of a winning proposal and price, in accordance with the required submission schedule, for a cost acceptable in view of the probable (financial) return to your organization.

PROPOSAL AUTHORIZATION	NUMBER	REVISION

TITLE

JOB

PERFORMANCE REQUIRED

ESTIMATED STARTING DATE	ESTIMATED DURATION

ESTIMATED BID PRICE	ESTIMATED SUBCONTRACT TO OTHERS $ %

IS JOB FUNDED?	WHAT IS FOLLOW-ON POTENTIAL?

ESTIMATED NEED FOR CAPITAL AND FACILITY EXPENSE IF JOB IS OBTAINED

CUSTOMER ORGANIZATION

KEY CUSTOMER PERSONNEL

CONTRACT FORM	SPECIAL CONSIDERATIONS	SECURITY CLASS

COMPETITION

COMPETITORS

COMPETITORS' STRENGTHS

SIGNIFICANCE TO COMPETITORS IF THEY LOSE

OTHER COMPETITOR WORK FOR CUSTOMER

OUR UNIQUE ADVANTAGES

PROPOSAL

WHAT IS TO BE SUBMITTED?

DUE DATE	PROPOSAL COST (DETAIL BELOW)

PROPOSAL MANAGER	OTHER KEY PROPOSAL PERSONNEL

ACTIVITY / EFFORT	PROJECT DEPT. (HOURS)	SUPPORT GROUP A (HOURS)	SUPPORT GROUP B (HOURS)	SUPPORT GROUP C (HOURS)	NONLABOR (DOLLARS)
PREPROPOSAL					
BIDDERS' CONFERENCE					
PROPOSAL PREPARATION					
CUSTOMER PRESENTATION					
CONTRACT NEGOTIATION					
OTHER					
TOTAL HOURS					
TOTAL COST					

FUNDING NEEDED	JAN	FEB	MAR	APR	MAY	JUN	JUL	AUG	SEP	OCT	NOV	DEC
MONTHLY												
CUMULATIVE												

APPROVALS

GROUP A	DATE	MARKETING MANAGER	DATE	VICE-PRESIDENT—FIN.	DATE
GROUP B	DATE	DIVISION CONTROLLER	DATE	EXEC. VICE-PRESIDENT	DATE
GROUP C	DATE	DIVISION MANAGER	DATE	PRESIDENT	DATE
PROPOSAL MANAGER	DATE	VICE-PRESIDENT—OPERAT.	DATE	CHAIRPERSON OF BOARD	DATE

FIGURE 3-4. *Typical proposal authorization form.*

Theme Fixation

Except in those very formal RFPs in which contact with the prospective customer is either prohibited or heavily constrained (for instance, some government RFPs), it is vital to spend time with your customer fixing the theme and central focus of your proposal effort. This includes clarifying which dimension(s) of the Triple Constraint should be emphasized.

There are several reasons that theme fixation is important. In the first place, many customers are in fact organizations, and the "customer" is comprised of people who view the contemplated undertaking in slightly different ways. It is necessary to understand these subtle differences and either harmonize them or deduce who has the most influence. This understanding is portrayed as the lower piece of bread in Figure 3-3. As such, it is all the knowledge from the preproposal work.

In addition, the customer's statement of the problem in the RFP may be imperfect or incomplete. Working with the customer—when this is permitted—can correct this problem as well as demonstrate your organization's competence. In the course of this dialogue with your customer, you will have opportunities to launch trial balloons representing your initial approach to the proposal. You will learn your customer's preference, which will permit you to adjust your thinking to produce a proposal that is more responsive to your customer's prejudices and predilections. The funnel in Figure 3-2 illustrates one way this can happen.

It is important for everybody in your organization to understand the chosen theme so their contributions are consistent with it. This theme will be used throughout your written proposal. It might be technical sophistication and elegance, early delivery, the fact that the unit you propose to furnish is a proven item, or that you have a team ready to put to work on the job. If everyone in your organization understands what the theme is, your proposal should be acceptable to your prospective customer.

Statement of Work

Content   The statement of work (SOW) must describe the job to be done. It should designate any specifications that will be applied. It should identify measurable, tangible, and verifiable acceptance criteria so there is no uncertainty whether the final item is in fact acceptable. It is also extremely helpful to know which dimension of the Triple Constraint is most important to the sponsor, which is second most important, and, by implication, which is least important. Unfortunately, some sponsors have not thought about this, and sometimes sponsors will not provide such a priority ranking. Nevertheless, if you can obtain it, it will help you make intelligent trade-offs if these are required. To reiterate what I said in the previous chapter, quality can be considered part of the performance dimension.

Specifications are crucial. Figure 3-5 shows factors that influence requirements, from which maximally satisfactory specifications for products or services must be derived. Specifications often require trade-offs, as illustrated in Figure 3-6. Other common trade-off examples include automobile carrying capacity can be substan-

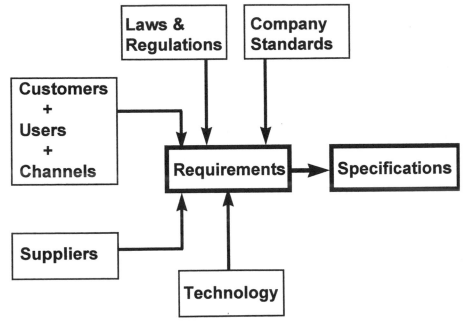

FIGURE 3-5. *Specifications should satisfy requirements, which are affected by diverse factors.*

FIGURE 3-6. *Example of a trade-off (between weight and cost) that must be made in the specification.*

TABLE 3-1 Illustrative Elements of Specifications

Quantitative	Qualitative
Size	Styling
Weight	Attractiveness
Speed	Aroma
Sensitivity	Taste
Resolution	Sound quality

tially enlarged only at the cost of poorer fuel mileage and photographic film speed (light sensitivity) can be improved only at the expense of resolution. One conscientious person's perception may differ from another conscientious person's perception, so these trade-offs must be explicitly resolved before team members begin to work at cross-purposes.

Specifications also have both quantitative and qualitative aspects, as shown in Table 3-1. Neither the proposal nor the project teams have much trouble understanding how to judge the quantitative ones, but they frequently get into trouble with the qualitative ones. Consider a product that must be "contemporary" or "attractive." Who decides, and on what basis? When forced to employ a qualitative specification, one must specify how satisfaction will be judged. It might be decided exclusively by the owner of the company, which is a common situation in a small or privately owned company. In a larger organization, many people may be involved. In either situation, the decision criteria must be understood, and judges of compliance must be kept informed.

As I said in Chapter 2, when it is not possible to be this precise at the earliest phase of a job, because the final product is not assuredly realistic or clearly attainable, it is important to undertake a two-phase project effort. The first phase, perhaps extending to a customer review, is quoted completely, but the whole job is quoted only approximately, in *Project goals must be specific, measurable, and attainable.* a nonbinding fashion. The whole job quote represents the proposer's best estimate of project requirements. The first-phase quotation, however, is firm (even if a cost reimbursable contract form is used) and includes sufficient effort to construct acceptance criteria for the rest of the job.

Clarification   After it is drafted, the SOW should be reviewed with the customer prior to further work on the proposal itself. Ambiguous words should be avoided, and it is desirable to be quantitative, using numbers and dimensions whenever possible.

An SOW such as "design and build an amplifier with greater gain than model L" or "explore the continental shelf for oil" can be ambiguous. The contractor might interpret the former as being satisfied by 10 percent greater gain. The *Make the SOW precise and measurable.* customer might expect the new amplifier to have 300 percent

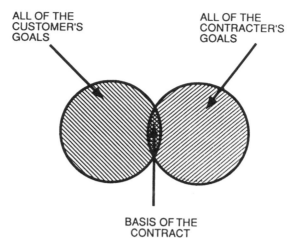

ALL OF THE
CUSTOMER'S
GOALS

ALL OF THE
CONTRACTER'S
GOALS

BASIS OF THE
CONTRACT

FIGURE 3-7. *The project manager must assure that other customer and contractor goals do not distract the people working on the project from accomplishing the contract requirement.*

This material originally appeared in a videotape presentation produced by the Association for Media-based Continuing Education for Engineers, Atlanta, Georgia.

> **Better is often the enemy of good enough.**

greater gain. This leads to what I call the "er trap." In too many situations, it is attractive to promise that a product will be faster, easier to use, smoother, lighter, roomier, sturdier, have a longer life, and so on without carefully determining what the "better" feature is truly worth to a user or buyer.

It is also important to be sure that extraneous requirements have been eliminated from the SOW. All too often "gold plating"—unnecessary customer or user requests or excessive elaboration or sophistication by your own personnel—creeps in. Figure 3-7 illustrates that the contract defines that narrow range of common interests that is to be dealt with in the project work. Other work—regardless of whether it is interesting or important—must be eliminated if it contributes unnecessary cost or adds too much time to the project schedule.

Plan

A proposal is a bridging step in the overall project process, being involved in both the "define" and "plan" activities of a project. The proposal will contain a plan; additionally, the act of writing the proposal forces the organization to think through and attempt to simulate the entire project.

Simulation   Simulation is used in many situations. Servomechanism engineers, for instance, will simulate on paper, and perhaps further using a digital or analog computer, the performance of a servomechanism before attempting to build even the breadboard. In doing this simulation, they are investigating how the servomechanism might perform if it is built according to certain specifications.

Civil or structural engineers will often examine the deformation of a building or a bridge to determine if it will have adequate strength if built according to the design drawings. Similar simulations occur in many other fields, such as aerodynamics, thermodynamics, and optical design. A major purpose of these simulations is to identify any potential problem areas in the prospective system before building it. If a design won't work on paper, it won't work when it's built.

Simulation aids prediction.

In the case of project planning, where the project plans are a simulation of how the project will be carried out, there are similar reasons for engaging in simulation. It is important to decide how to establish a price for the proposed work. A detailed simulation, that is, a plan, makes it more likely that the proposed or bid price will be sufficient. If the plan has been thoroughly prepared, it will also convince the customer that your organization understands the proposed job, which helps in negotiating your contract. If the project schedule is unrealistic on paper, the project won't be completed on time. Project management software, which I discuss in Part 2 and also in Chapter 26, can be used quite effectively for this schedule and budget simulation work.

If the schedule doesn't make sense on paper, the project will be completed late.

The Triple Constraint   Because the Triple Constraint is so important to planning and planning is so important to the proposal process, Part 2 of this book is devoted to planning. But it is important to understand that the project plan devised during the proposal process and presented in the proposal is a plan to satisfy the Triple Constraint. You use the work breakdown structure (WBS) to describe your approach to the performance dimension. Many computer project management software packages have a WBS capability. You use network diagrams or in some cases a bar chart, in which each activity corresponds to a WBS element, to describe your approach to the time dimension. You use a complete cost breakdown for each activity to describe your approach to the money dimension and defend your price.

These three planning elements are best prepared in the order presented. First, the WBS is used to describe those things that will be undertaken to satisfy the performance specification. After that is complete, it is possible to prepare a network diagram for each of these designated elements in the WBS. Initially, each of these items should be estimated in a "natural" time frame; then these activities' logical relationships to each other can be established. If, as usually happens, this produces an unacceptably long program, it is then important to decide which activities will be scheduled to be completed in periods shorter than the "natural" time. That is, some of the project activities must be carried out faster than is desirable. After agreeing to shorten the schedule, prepare cost estimates for each activity. Note that it is not desirable to prepare the cost estimates prior to determining the time to be allowed for a given activity. These issues will be treated in more detail in later chapters.

The plan has three dimensions.

Checklists   Checklists are designed to help assure that nothing that will have to be dealt with during the course of the job has been forgotten or omitted from consideration in the proposal. A checklist should contain such items as shown in Appendix 3. These checklists are not exhaustive and may not contain the most significant or most important items. They are meant to suggest the kinds of items that might appear in a checklist.

Take advantage of experience.

The best way to develop a checklist is to create your own over a period of years. One way is simply to enter items on file cards whenever they occur to you during the course of project work. You may later sort out these cards alphabetically, but time phase, or by some other logical method. Having developed a checklist from your own experience, you will perform better on future projects because you will not forget items that are likely to be significant. Thus, you will consider their impact on a project during the proposal phase; they will not emerge as unexpected developments during performance of the project.

On a new proposal or project, think about it critically before consulting your checklist. If you turn to the checklist first, you tend to develop "blinders" to new crucial factors unique to the current job.

Adjustments

Adjustments are often required after a proposal has been partially prepared. Perhaps someone discovers that two departments contributing to the proposal have duplicated their efforts or have made differing assumptions about some significant item. Or perhaps someone discovers new information or corrects some oversight.

When an adjustment is required, all participants must join in deciding how to make it. The proposal manager should not assume this responsibility. Two benefits accrue from participants making the adjustment. First, the experts are considering the problem and presumably making the most sensible adjustment. Second, having contributed to the adjustment, other participants gain a sense of involvement in the decision and tend to perform the job better when the proposal has been converted into a project undertaking.

Approval

As with the initiation of the proposal effort, the conclusion of the proposal requires a managerial action within an organization. There will typically be a sign-off control sheet (Figure 3-8). Normally, each organization has a procedure that specifies the signature authority of given managerial levels, and such a procedure indicates which managers or officers must sign the control sheet signifying their approval for proposal submission. The sign-off control sheet must contain a brief description of the Triple Constraint contained in the proposal document being submitted. The sign-off control sheet is retained in the proposing organization's files and is not submitted to the customer.

PROPOSAL SUBMISSION APPROVAL	COMPANY PRIVATE	
PROPOSAL TITLE	NUMBER	
CUSTOMER	CONTRACT FORM	
SUMMARY STATEMENT OF WORK		
SCHEDULE FOR JOB		
COST	FEE	TOTAL BID PRICE
DOCUMENTS, REPORTS, MODELS, ETC., SUBMITTED		
SUMMARY OF OUR RISKS		
KEY PEOPLE PROMISED		
FINANCIAL COMMIITMENTS REQUIRED		
WARRANTY		
ACCEPTANCE CRITERIA		
REMARKS		
MARKETING MANAGER DATE	VICE-PRESIDENT—FIN. DATE	
DIVISION CONTROLLER DATE	EXEC. VICE-PRESIDENT DATE	
DIVISION MANAGER DATE	PRESIDENT DATE	
VICE-PRESIDENT—OPERAT. DATE	CHAIRPERSON DATE	

FIGURE 3-8. Typical proposal submission approval form.

Plan ahead for the proposal approval.

It is important not to take the approval of senior managers for granted. Therefore, it is important to give these people timely briefings throughout the proposal preparation effort as to the scope of the proposal and the nature of the resources to be committed to the resulting project. Although the proposal authorization document constitutes one such involvement of senior management, it alone will not suffice. The number and frequency of such briefings during proposal preparation depend on the organization, its rules and procedures, and the proposal manager's good judgment.

In the case of proposals to government customers, there is usually a requirement for an officer to sign a certificate of current pricing or similar form. This is intended to assure the government that the proposed price is calculated in conformance with current laws and procurement regulations. Because the signer of this certificate is subject to substantial personal penalties if the certificate is inaccurate, this officer will require a careful review of your proposal SOW and cost estimate.

Submission

The time comes to submit the proposal to the designated recipient, who may require that it bear a postmark by a certain date or be received at a given office by a specified date stamped by a particular time. Such standards are overriding and must be complied with.

As a practical matter, proposals should normally be submitted in three sections, as illustrated in Figure 3-9. Each of these sections might actually be a separate volume, and, in some cases, the technical and management portions as well as each appendix might be separate volumes. The exact content of each part may be stipulated by the RFP. If you generate a proposal on your own initiative, you may have more freedom to include what you want in locations that you judge will be best for the intended recipient.

Postsubmission

Mailing or delivering the proposal is not the end of a winning proposal effort. Postsubmission work is analogous to the upper piece of bread in the proposal sandwich. At the very least, the winning organization must negotiate a contract with the customer. Sometimes several proposing organizations are deemed qualified, and negotiations are carried out with two or more of them prior to selecting the winning contractor.

More work—and expense—follows submission.

In many proposal situations, the negotiation phase is preceded by a presentation to the customer. Such a presentation may be elaborate, requiring special graphics and models, and may entail extensive time and effort.

To recapitulate, the following are steps in the proposal process: Establish the organization's business strategy. Understand the organization's resources. Get the authorization (bid/no bid). Make the preproposal effort. Receive

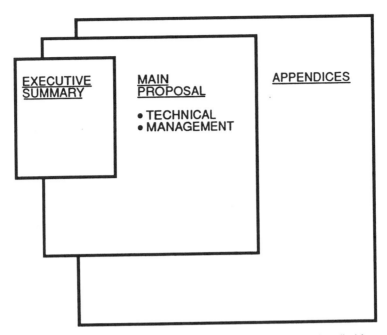

FIGURE 3-9. *The submitted proposal should be presented in an easily handled format.*

This material originally appeared in a videotape presentation produced by the Association for Media-based Continuing Education for Engineers, Atlanta, Georgia.

the RFP. Attend the bidder's conference. Fix the theme. Prepare the statement of work. Plan the job. Adjust the proposal. Approve the proposal. Submit the proposal. Present the proposal to the customer. Negotiate the contract. The proposal contains a statement of work, which is the basic project definition. To prepare the proposal, it is necessary to do some, but not all, of the project planning. Much of this planning is commonly included in the proposal.

TYPICAL PROBLEMS

There are both practical and people problems with proposals for projects. The first practical problem encountered in proposing winning projects is attempting to do virtually the entire job during the proposal. That is, in trying to prepare a solid proposal, you spend too much time working through the plan for the project. Sometimes this includes doing preliminary engineering, modeling, or program coding. You can overcome this by recognizing that risk must be balanced and planning only enough to reduce the project's uncertainty to an acceptable level.

(continued)

A related problem is inadequate project planning in the proposal. The solution here is to keep planning until it becomes too time-consuming. This is obviously a judgmental issue for which personal experience must provide guidance. Another problem is the last-minute rush to complete the proposal in time to submit it. The solution to this problem is to have a proposal preparation schedule and adhere to it. Another frequent problem is a poorly crafted specification, regardless of whether the proposal is initiated in response to an RFP or self-initiated. The solution is to be explicit about both the quantitative elements and the means by which the qualitative elements will be judged.

The principal problem with people is getting the workers who will subsequently perform the work on the project to contribute to the proposal. The ideal at which to aim—seldom completely achieved in practice—is to have all the people who will be key project participants describe, plan, and estimate the work they will later do. This is the Golden Rule discussed in Chapter 5.

MATERIALS STUDY PROJECT

After being approached by NERESCO, you reviewed their work with SUPROMAC management. SUPROMAC has done prior work for NERESCO and the materials study project is deemed a lead to a still larger effort; so SUPROMAC is anxious to propose on the work, and a proposal is authorized (see Figure 3-10). Your discussions with NERESCO about different initial perceptions of the desired Triple Constraint have led to an agreement to study only two materials, thus conforming to their wish for a six-month project and staying within their budget of $500,000. You have also agreed on which two materials will be studied. But they want a theoretical study included to reduce the risk resulting from omitting the study of the other four materials you felt should be studied.

Thus, you can now write the proposal. The SOW will name the two materials, the specific tests to be performed, and the kinds and amount of data to be delivered. It will also contain a detailed schedule of work and cost breakdown.

HIGHLIGHTS

- Organizations must filter out losing projects.
- Winning projects arise from good proposals, thoughtfully initiated to be consistent with the organization's goals.
- Four issues involved in the decision to bid on a proposal opportunity are the nature of the requirement, the value of the project, the organization's response ability, and its ability to win.

PROPOSAL AUTHORIZATION		NUMBER *1728*	REVISION *ORIG.*

TITLE *MATERIALS STUDY*

<table>
<tr><td rowspan="10">JOB</td><td colspan="2">PERFORMANCE REQUIRED
STUDY 2 MATERIALS</td></tr>
<tr><td>ESTIMATED STARTING DATE
JAN 01</td><td>ESTIMATED DURATION 6 MONTHS</td></tr>
<tr><td>ESTIMATED BID PRICE
$500,000.00</em></td><td>ESTIMATED SUBCONTRACT TO OTHERS
$ $5000. % 1%</td></tr>
<tr><td>IS JOB FUNDED? YES</td><td>WHAT IS FOLLOW-ON POTENTIAL?
VERY LARGE ($5M?)</td></tr>
<tr><td colspan="2">ESTIMATED NEED FOR CAPITAL AND FACILITY EXPENSE IF JOB IS OBTAINED
USE EXISTING EQUIPMENT & SPACE</td></tr>
<tr><td colspan="2">CUSTOMER ORGANIZATION
NERESCO</td></tr>
<tr><td colspan="2">KEY CUSTOMER PERSONNEL
KIM DRIVER</td></tr>
<tr><td colspan="2">CONTRACT FORM FP   SPECIAL CONSIDERATIONS —   SECURITY CLASS UNCL</td></tr>
</table>

<table>
<tr><td rowspan="5">COMPETITION</td><td>COMPETITORS UNCERTAIN — PROBABLY SOLE SOURCE</td></tr>
<tr><td>COMPETITORS' STRENGTHS
UNKNOWN</td></tr>
<tr><td>SIGNIFICANCE TO COMPETITORS IF THEY LOSE
UNKNOWN</td></tr>
<tr><td>OTHER COMPETITOR WORK FOR CUSTOMER
UNKNOWN</td></tr>
<tr><td>OUR UNIQUE ADVANTAGES LOT'S OF DISCUSSION, CAN DO IT, MAYBE SOLE.</td></tr>
</table>

PROPOSAL

WHAT IS TO BE SUBMITTED? *PROPOSAL*

DUE DATE *15 NOV 91*   PROPOSAL COST (DETAIL BELOW)

PROPOSAL MANAGER *MEL CHASE*   OTHER KEY PROPOSAL PERSONNEL *CHERYL WILLIAMS*

ACTIVITY / EFFORT	PROJECT DEPT. (HOURS)	SUPPORT GROUP A (HOURS)	SUPPORT GROUP B (HOURS)	SUPPORT GROUP C (HOURS)	NONLABOR (DOLLARS)
PREPROPOSAL	DONE				
BIDDERS' CONFERENCE	—				
PROPOSAL PREPARATION	20	8			
CUSTOMER PRESENTATION	4	4			
CONTRACT NEGOTIATION	8				
OTHER					
TOTAL HOURS	32	12			✕
TOTAL COST	800.	300.			

FUNDING NEEDED	JAN	FEB	MAR	APR	MAY	JUN	JUL	AUG	SEP	OCT	NOV	DEC
MONTHLY											0.7	0.4
CUMULATIVE											0.7	1.1

<table>
<tr><td rowspan="4">APPROVALS</td><td>GROUP A   DATE</td><td>MARKETING MANAGER   DATE</td><td>VICE-PRESIDENT—FIN   DATE</td></tr>
<tr><td>GROUP B   DATE</td><td>DIVISION CONTROLLER   DATE</td><td>EXEC VICE-PRESIDENT   DATE</td></tr>
<tr><td>GROUP C   DATE</td><td>DIVISION MANAGER   DATE</td><td>PRESIDENT   DATE</td></tr>
<tr><td>PROPOSAL MANAGER   DATE</td><td>VICE-PRESIDENT—OPERAT   DATE</td><td>CHAIRPERSON OF BOARD   DATE</td></tr>
</table>

FIGURE 3-10. Typical proposal authorization form.

- The proposal defines what the project will accomplish.
- A checklist may help you avoid overlooking required work.
- The proposal process includes authorization, theme selection, SOW preparation, plan development, adjustment, approval, submission, and postsubmission follow-up.
- A specification, which can include both quantitative and qualitative elements, involves trade-offs that must be resolved.
- Project proposals describe the Triple Constraint with a work breakdown structure, activity network diagram or bar chart, and cost estimate for each activity, which then serve as a project plan.

4

Negotiations and Contracts

Negotiations between the customer and the contractor convert the final definition of work into a contract. This chapter emphasizes the importance of appropriate contract forms and the necessity to start the proposal process with attention to this issue. It also discusses the special case of projects involving a foreign customer.

NEGOTIATING THE CONTRACT

Preproposal Negotiations

Contract negotiation really begins in the proposal phase because the expected contract form must be consistent with the job to be undertaken. If there is any reason to believe the customer will require a fixed price contract (Table 4-1), for instance, and the job calls for a major technological

Make a proposal only if the contract will be acceptable.

advance you are not certain you can achieve, it would not be prudent to continue in the preproposal and proposal effort. Hence, one objective of preproposal activity and discussions with a customer prior to major proposal expense is to assure that the contract form they intend to issue is consistent with the contract form your company or organization is willing to negotiate considering the work to be undertaken. In addition, negotiations are designed further to improve the likelihood of the customer organization and contracting organization having the same perception of the job.

Customer and Contractor Perceptions

The perceptions of both contracting parties must be harmonious. It is always possible for an RFP or the responding proposal not to be completely clear. If the preproposal

TABLE 4-1   Common Contractual Forms

Abbreviation	Definition
FFP	Firm fixed price—the price and fee are predetermined and do not depend on cost
FP	Fixed price—same as FFP
CPFF	Cost plus fixed fee—the customer agrees to reimburse the contractor's actual costs, regardless of amount, and pay a negotiated fixed fee independent of the actual costs
CPIF	Cost plus incentive fee—similar to CPFF except the fee is not preset or fixed but rather depends on some specified incentive
T & M	Time and material—the customer agrees to pay the contractor for all time and material used on the project, including a fee as a percentage of all project costs

process did not remove these potential misperceptions, the negotiation process offers the last opportunity to do so. Specifically, the final deliverables—oral, written, or a more tangible product—must be well defined, and the criteria for measuring or judging acceptance and completion must be straightforward. In some instances, acceptance occurs only after the product has left the manufacturer's possession, as, for instance, when a satellite must perform to some specified level in orbit or the actual operating cost of a piece of machinery must be less than a stipulated amount when that machinery is operating in a factory.

Contractual Forms

At the simplest level, there are no contracts. This is typical of self-financed efforts such as a project to paint your own house. This situation also prevails in efforts supported by your own organization, such as development of a new product's production line for use within your own company.

When one organization enters into a contract with another organization, there is a variety of possible contractual forms (see Table 4-1). In the first of these, very common in commercial situations, the contract is a so-called fixed price (FP) or firm fixed price (FFP) contract. This has the lowest financial risk to the customer because the maximum financial obligation is specified; conversely, the FP form has the highest financial risk for the contracting organization but offers the highest potential reward if the estimated costs can be underrun. This is illustrated in Figure 4-1. There are variations on this in which the price allows for escalation or redetermination due to some set of factors, such as inflation. Or there may be an FP contract with an incentive fee based on some performance aspect, perhaps early delivery.

Another class of contractual arrangement includes those in some way cost reimbursable. Here the customer bears an obligation to reimburse the contractor for all costs incurred; so the customer has a high financial risk and the contractor has a correspondingly low risk. Typical contracts of this sort are cost plus a fixed fee

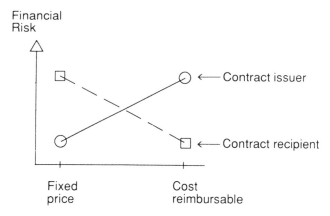

FIGURE 4-1. *The contract issuer and contract recipient have complementary financial risks.*

(CPFF) or cost plus an incentive fee (CPIF). Time and material (T & M and rate of effort) contracts are also a form of cost reimbursable contract.

Contracts entered into with the U.S. Department of Defense are governed by complex regulations, which are special versions of federal procurement regulations. In the case of the FP contracts, the government is never obligated to pay more than the specified amount. However, if the contractor performs very well and manages to underrun the cost budget

The contract form puts risk on either the customer or the contractor.

substantially, the government has the right to reduce the amount paid to the contractor below the price specified in the contract. Thus, FP contracts are a one-way street to the government's advantage.

An FP contract is preferable when there is virtually no uncertainty about your ability to satisfy the performance specification on schedule and for, at most, your estimated costs. Such situations typically occur when you have done essentially the identical or very similar project work previously and you have the appropriate human and physical resources available. If these conditions are not satisfied, the FP contract is a financial gamble and should be undertaken only if the risk is acceptable and the prospective reward is commensurate with it.

Negotiations

In a typical negotiation, the customer attempts to increase the performance specification while reducing the schedule and budgeted cost. If it is a competitive solicitation, the customer will often play off one prospective contractor against another to try to maximize his or her apparent benefit. Therefore, one should expect the customer to behave in this way.

Expect the negotiated contract price to be less than your proposed price.

In addition, negotiators should understand clearly how far, if at all, their management is prepared to deviate from the terms and conditions offered in the submitted proposal it approved. Second, good planning aids negotiators in that there is a complete work breakdown structure, with an attendant activity schedule and cost estimate for each element of it. Project management software can be helpful in planning "what-if" trade-offs. These help negotiators understand the job being negotiated and can usually help explain and/or defend it to the prospective customer.

In preparing for negotiations, it is frequently desirable for your organization to conduct a trial run with someone or some group playing the customer. In short, be well prepared and know your minimum acceptable position. Also, because only one negotiating team member should talk at once, you can rehearse who will respond to particular issues if they are raised.

After two dimensions of the Triple Constraint have been specified, the third can be determined.

During actual negotiations, it is usually to your advantage first to define the job (the detailed statement of work, specifications, and test criteria) and the schedule. After that, you can negotiate the exact contract form, including any detailed terms and conditions, and the final price. The effective negotiator always "horse trades" and never makes a unilateral concession.

Nevertheless, there is often give and take in the negotiation process, and some changes may be agreed to. Whenever there is a change in one element of the Triple Constraint, there must be changes in other elements. For instance, a customer may offer to provide customer furnished equipment (CFE) to reduce the expected cost of some activity within the proposed project. In the case of the government, this is called government furnished equipment (GFE). When this occurs, language on the performance axis must be changed to indicate the performance specifications the CFE must meet and the schedule of CFE provision must be stipulated. When both these things have been accomplished, it is possible to agree to substitute the CFE for contractor procured items and offer a reduction in the proposed schedule or budget. There is an infinity of possible trade-offs, including, for instance, deciding who is responsible for laboratory space and equipment, physical security, test equipment, parts storage and control, and so on.

A negotiated change in one dimension on the Triple Constraint should be accompanied by changes in the other dimensions.

When the negotiations are concluded, the binding direction on the project is that specified in the contract. Only contract change notices suitably signed and agreed to by both parties can permit changes. Such changes should not occur outside the contracting mechanism, despite agreements reached by members of the contractor's and customer's support teams—for instance, when two engineers meet and agree that a new microcircuit would be a desirable item to include. Verbal redirection, which can easily occur because of the many individuals involved, is not binding until the contract is amended, as shown in Figure 4-2. Renegotiations such as these require the same amount of planning and preparation as the original negotiations. It is not uncommon to have several renegotiations during a long or complex project.

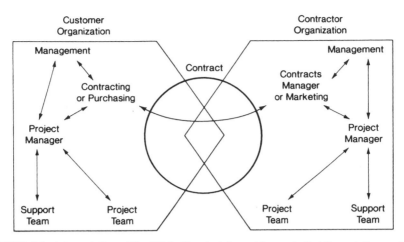

FIGURE 4-2. *Interpretation of the Triple Constraint must be controlled through the contract.*

Legal Aspects

There is a myriad of legal aspects in project contracts. Discussions with suitable people in your organization are often helpful.

In the case of U.S. government contracts, there is a host of special regulations. A typical government contract will include the following:

1. Customer's name and address and specification of authorized signatures
2. Contractor's name and address and specification of authorized signatures
3. Statement of supplies (items), services, and prices (statement of work)
4. Preservation, packaging, and packing instructions
5. Delivery or performance period
6. Inspection and acceptance terms
7. Contract administration data
8. Special provisions (funding limitations or customer furnished equipment)
9. General provisions (reference to federal procurement regulations, overtime payment terms, or similar) such that, for example, manufacturers have been obligated to pay criminal and civil penalties for violations of federal regulations
10. Patent terms, conditions, and ownership rights (if not covered in items 8 and 9)
11. List of required documentation

Despite all this intricate detail, it is still possible to find ambiguity. One company sued to invalidate the government's right to contract for delivery of systems for which specific congressional payment authority was not yet enacted. Although this is largely a contractor's ploy to renegotiate for a higher contract price, it demonstrates how seemingly clear contracts can be disputed.

Misrepresentation of costs is illegal and can produce serious consequences. There can be other problems with government contracts. There are many instances when government contractors (federal, state, or municipal) are vehemently criticized for allegedly inadequate project performance, and this is often widely reported in the media. All major newspapers carried extensive articles reporting on the chairman of a congressional committee's criticism of McDonnell Douglas for its work on the C-17. Worse outcomes also occur. The U.S. Navy canceled a major contract for a new antisubmarine airplane (the P-7A) with Lockheed in mid-1990, alleging that Lockheed's performance was inadequate. With counterclaims, attorneys will now be involved, perhaps for a long time, and Lockheed will suffer substantial financial consequences until the matter is finally settled (*New York Times,* July 21, 1990, p. 30L). In addition, in the case of fixed price contracts, failure to deliver can expose the contractor to very serious cost penalties far beyond absorbing the cost of its own effort. Such penalties may not only include termination but may obligate the contractor to pay the costs of a substitute the government hires to deliver after the contract is in default.

If any patents are obtained in the course of the project, the government may own them. In addition, there are frequently highly proscribed procedures specifying how much of the work throughout the project must be performed. Costs are frequently subject to audit when the job is completed, and the amount finally paid to the contractor is frequently reduced.

Understand the legal implications. In the case of commercial contracts, restrictions are far less one-sided. In construction projects, for example, there may be clauses about what will be done if unanticipated rocks are encountered below ground, delays are caused by unusual weather, or similar situations occur. Force majeure (that is, an act of God) may permit a contractor to escape any penalty for failure to deliver on time. In addition, the suitably drawn contract will specify that disputes are to be settled by arbitration rather than by the courts. Arbitration is far simpler, quicker, and less costly to both parties. Nevertheless, antitrust regulations and many other laws limit the kinds of commercial arrangements into which two companies may enter.

INTERNATIONAL PROJECTS

International projects are not fundamentally different from domestic projects. In practice, however, the travel lure of a remote and apparently salubrious destination seems to affect human judgment. Furthermore, international projects expose a contracting organization to special problems related to language, currency, and unfamiliar business practices.

Remoteness

The other party in an international project is typically located in some remote area of the world. The far-off hills often look green. When two project opportunities are available to a company or organization at the same time, the one that

Recognize the travel lure.

originates in Zurich will receive far more attention than the one that originates in Cleveland, and this is independent of the intrinsic merits of the project opportunity. This is the "travel lure," and it is a problem you must recognize and identify for what it is.

There are occasions, however, when you should undertake a project involving a foreign customer or partner. When this is the case, it is desirable to get more information about the foreign organization or the business in which it is engaged. If your organization has an office in the country, that is the best source of information. The foreign organization's country may have an information office in the United States, which may be able to provide useful information. The United States embassy or trade centers in the foreign country may have information, as may the U.S. Department of Commerce or the World Bank.

Having gathered all the information you can without leaving the United States, you may then have to succumb to the travel lure and go to the foreign country to discuss the project or negotiate. It is generally prudent to go several days before the initial meeting to compensate for jet lag.

Business Practices

The business practices governing project performance as well as negotiations and discussions leading to the project will typically be those dictated by the customer's country. For instance, Japanese tend to discuss a project at great length to try to achieve consensus. When this has been done, a handshake or verbal agreement is binding.

Staffing requirements will frequently require that some of the work be reserved for the customer's nationals. Sometimes nationals of a third country are stipulated or prohibited, for a variety of reasons. Permits or "red tape" enter into business practices, and these will often absorb far more time than expected. In some cases, export or import controls can impose severe restrictions; so it is always important to understand these issues early.

Laws and Regulations

These are numerous laws and regulations, unfamiliar to the vast majority, that apply to international business. Limitations on the export of technology from the United States make headlines periodically when someone or some company is charged with a violation. Such violations typically involve shipment (or indirect reshipment) of some restricted high-technology product to unfriendly nations.

In other situations, even more arcane restrictions, such as the Export Administration Act, may snag the unwary project manager. In one case, a United States congressman allegedly broke a law prohibiting U.S. citizens or companies from aiding the Arab boycott of businesses tied to Israel merely by writing a letter to someone in Kuwait. Thus, if you are contemplating an international project, be sure to have competent legal guidance on what you can and cannot do.

Language

Be prepared for unfamiliar customs, language, and currency.

Fortunately for Americans and people from other English-speaking nations, most contracts are written in English. In some cases, however, a customer's language will prevail, and the contract will be written in it. If you are using a translation, it is prudent to have several prepared because the original host language document is the controlling document and interpreters are likely to interpret it differently. If you find wording differences, you can explore their significance with a language expert.

Price

As with language, the customer's currency may be the stipulated medium of financial settlement. Because currency rates fluctuate, it may be desirable to insure your company against these fluctuations. This is done with a hedge contract. Alternatively, you may use letters of credit to satisfy the future payment obligations. In any event, your price must allow for the extra expense of doing business at a great distance.

TYPICAL PROBLEMS

The most common negotiation problem is deciding how to cope with the inevitable price squeeze the customer will try to inflict. Several things help:

1. A good plan, well explained
2. A clear understanding of where you have inserted a negotiating "cushion" or "fat" (as distinct from contingency, which is discussed in Chapter 11) and a negotiating plan on how to horse trade it
3. Management guidance, or clearance, on how much you can give up
4. A reputation for having met prior commitments

Finally, if you are forced to surrender schedule or price, be sure to try to alter the performance specifications or obtain CFE.

MATERIALS STUDY PROJECT

Preproposal discussions between SUPROMAC and NERESCO have narrowed the scope of negotiations considerably. Because SUPROMAC believes this project will lead to more work later, it is not interested in taking a tough and unbending posture. Nevertheless, SUPROMAC might first try to justify a price in excess of $500,000 because of the theoretical study NERESCO substituted for the four materials omitted from the study.

On the presumption that an FP contract is jointly agreed to, SUPROMAC can attempt to gain favorable detailed terms and conditions. These might include progress payments and a degree of informality in the required reports.

HIGHLIGHTS

- A proposal should be made only if a reasonable contract can be negotiated.
- There are several contractual forms, including fixed price, firm fixed price, cost plus fixed fee, cost plus incentive fee, and time and material.
- Both parties must be prepared to make concessions during negotiations.
- Government contracts include numerous special regulations.
- International projects introduce special problems, such as unfamiliar business practices, laws and regulations, language, currency, and distance.

Part 2

Planning a Project

5

Why and How to Plan
a Project

The planning activity for the management of a project is crucial. Plans are the simulation of a project, comprising the written description of how the Triple Constraint will be satisfied. Therefore, project plans are really three plans: one for the performance dimension (the work breakdown structure), one for the schedule dimension (preferably a network diagram but occasionally a milestone listing or bar chart), and one for the cost dimension (a financial estimate). This chapter covers plans in general, reiterates the need for plans, describes how these three kinds of plans are made, reviews several planning issues, and discusses "the Plan," a document or series of documents embodying the project's planning agreements.

PLANNING

In broadest generality, plans depend on three factors:

1. Knowing where you are now (or will be when whatever is being planned for will start)
2. Knowing where you want to get
3. Defining which way you will get from where you are to where you want to be

These factors are illustrated in Figure 5-1. The old saying, "When you don't know where you want to go, any road will get you there," is true; you can have a plan only if you have a destination in mind.

Many company sponsored projects are initiated because of the organization's long-range plan. Thus, plans are frequently hierarchical, with short-range plans established within the context of long-term plans. For instance, project task plans are components of the overall project plan. As mentioned in Chapter 1, task plans are components of project plans and so on. In addition, planning is iterative, so project plans must be revised when other plans change. For example, when the long-range

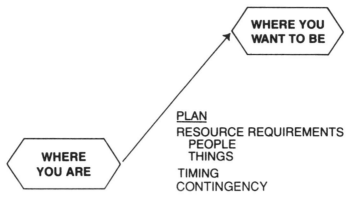

FIGURE 5-1. *Planning.*

plan covers five or ten years, changes obviously occur, priorities must be altered, and projects are added or canceled in response to the dynamic environment.

An effective project plan accomplishes the following tasks:

1. Identifies everything required to successfully complete the project
2. Contains a schedule for the timing of these tasks and related milestones
3. Defines the required resources, assuring that these will be available at the appropriate time, and reflects the participation of these resources and their management
4. Has a cost budget for each task
5. Includes suitable contingency
6. Is credible to both the planned performers and management

THE NEED FOR PLANS

Plans aid coordination and communication, provide a basis for monitoring, are often required to satisfy requirements, and help avoid problems.

Coordination and Communication

Most projects involve more than one person. Typically, a technical expert is asked to perform in the area of his or her expertise. For instance, a carpenter works on wood framing, not the plumbing or electrical wiring in a construction project. Similarly, an expert on electronic circuit design works on the electronic circuit design task, not on the optical design task for an electro-optical system. The project plan is a way to inform everyone on the project what is expected of him and her and what others will be doing. Plans are a vehicle to delegate portions of the Triple Constraint down to the lowest (task or subtask) reporting level.

Let others plan their work.

If the people responsible for these tasks also participate in making the plans, they will have an added impetus to adhere to them. Thus, there is a Golden Rule for planning:

> Get the persons who will do the work to plan the work.
>
> · They should know more about it than anyone else.
> · It's *their* task, not *yours*.

A corollary to the Golden Rule: No job is too hard for the person who does not have to do it, but persons who plan their own tasks are more likely to be able to do them.

Your project plans matter. Even if your project can be performed in your office, other people in the organization (for instance, your boss) will want to know where your project is headed, what you are doing, and for how long you will be doing it. Thus, project plans constitute an important communication and coordination document and may motivate people to perform better.

Basis for Monitoring

Plans are also the basis of your project monitoring activity (see Figure 5-2). It is a characteristic of projects that they do not go in accordance with plan. What you do not know when you start is where and how your project will go off plan. Deviations from plan, detected by monitoring progress, constitute your early

Projects do not follow their plan.

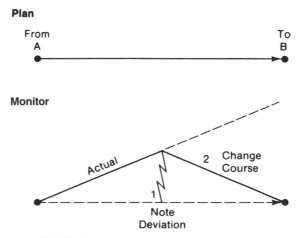

FIGURE 5-2. *Plans are the basis for monitoring.*

This material originally appeared in a videotape presentation produced by the Association for Media-based Continuing Education for Engineers, Atlanta, Georgia.

warning signal during project performance that there are problems to be resolved. That should cause replanning to occur.

When an airplane takes off from Los Angeles headed toward Honolulu, there is a planned course to be followed. During the flight, the navigator makes measurements of the plane's position, noting whether the plan being followed has deviated from that planned. If there is a deviation, the pilot changes course after the navigator notes the deviation so the plane can get to Honolulu. If no flight path corrections are made, the plane might have to land in the middle of the Pacific Ocean, which would be as bad for the plane as a similar outcome would be for a project that deviated too far from its intended Triple Constraint point.

Plans are a detailed description, formulated before the project is carried out, for accomplishing its various aspects. Deviations may indicate that the project will not reach its intended destination.

> **There are plans for all three dimensions of the Triple Constraint.**

Figure 5-3 illustrates how all the techniques covered in Part 2 of this book are related. The work breakdown structure (WBS) leads to a network diagram. Each activity or task on the network is then time phased and checked for resource allocation to be certain it is consistent. When this is done, it is possible (although not necessary) to construct a schedule bar chart indicating the time frame for each activity or task of the entire project. The costs for these tasks are then estimated, and when work is being performed on them, costs are reported back by specific task in each expense category. Part 4 of this book discusses how this detailed variance information can be used to help manage the project.

Requirement Satisfaction

Plans are sometimes created merely to satisfy requirements imposed by others, perhaps a customer or your boss. In such a situation, plans are often created under duress rather than because they are perceived to be valuable, even essential, in achieving project objectives.

Plans so created are frequently not followed. All too often they are generated and then discarded because they were prepared only to meet the requirement to prepare a plan. When there is such a requirement and the plans are prepared slavishly rather than thoughtfully, it is a waste of time for the preparer and the reader.

Problem Avoidance

Project management is sometimes a race with disaster. This is frequently the case for less experienced project managers. All too often the last crisis has scarcely been resolved before the current one begins, and then the project manager is too busy to anticipate and try to head off the next one.

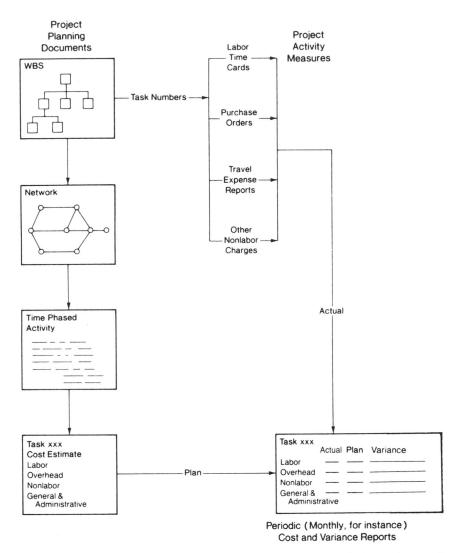

FIGURE 5-3. *Overview of how project planning documents are used in the project monitoring phase.*

A good plan helps you avoid problems during project performance (but plans cannot prevent problems). Consider the following example of a schedule and cost problem. Your project requires a final report. You assume it will require sixty pages and twenty figures, and the technical documentation group, which will prepare the report, agrees to do it in one week for $1,200. When you later ask them to prepare a one-hundred-twenty-page documentation report with forty figures, you will be told you will get it two weeks later and it will

Planning is crucial; a good plan is essential.

cost $2,400. Obviously, you cannot "plan" for a two-hundred-page documentation report with one hundred figures because the price in your proposal may be too high for you to win the job. You must make the best plans and estimates you possibly can and then try to adhere to them. For instance, as you get to the final phase of the project and must prepare the report, assign writing to the participants in such a way that they all clearly understand the planning goal. Constrain the writing efforts to adhere to plans. If you do this, the report should be approximately sixty pages and twenty figures.

PLANNING ISSUES

Uncertainty and Risk

Plans relate to future events. That is, your plans are a simulation of how things will occur in the future. There are necessarily uncertainties about the future, some of which may be somewhat predictable and thus partially controllable, but many of which are unpredictable.

Your objective is to be as confident as possible about those things that are predictable. You would also like to be fully confident that you will successfully achieve the Triple Constraint. But there will always be uncertainties, unanticipated tasks and outcomes, unexpected options, and unfortunate mistakes; so total confidence in complete success is naive. The best you can do is work toward all the objectives you can clearly perceive and recognize that there will be many totally unexpected developments that will challenge you and your team.

Use checklists to reduce uncertainties.

You can reduce (but not eliminate) these predictable uncertainties by using checklists, thoroughly discussing the plans with experts, and involving your entire team. Nevertheless, uncertainties will remain because there are always unpredictable factors when you are doing something new. Allowance for these unknowns can be made by inserting contingency in your plans, but the unknowns cannot be eliminated. For instance, thorough plans cannot prevent bad weather or strikes from delaying a construction project or eliminate cost changes due to currency rate fluctuations. Plans can be no better than your present understanding. If you have done something similar before, you can plan it better than if it is entirely new to you and your team. For instance, previous experience with an organic polymer formulation is not terribly helpful for planning a materials study project.

Assumptions, such as which people will be able to work on your project, are involved in planning. The plan for the materials study project looks a lot different when a senior mechanical engineer will do the work than when a junior electrical or chemical engineer will. Because assumptions are involved in your planning, it is important to include contingencies, which I discuss in Chapter 11. Good plans are quantitative rather than qualitative and as precise as possible.

A Choice Between Options

In preparing plans, as in carrying out project work, you are frequently confronted with options. These choices may include program management options, product quality standards, extent of subcontracting to be undertaken, and so on. Your plan may be considered the record of your choices between these options and will normally depend on how much risk you are willing to take or how much contingency allowance is included in your plans. Figure 5-4 illustrates this kind of choice. It shows three possible activity sequences. Which you choose depends on what is important to you, in this case, a short schedule (option 2) or the higher assurance of instrument accuracy (option 1). Another possibility (option 3), with two activities being performed simultaneously (in "parallel," as it is sometimes called), requires two people (one to verify, one to collect); the other two options require only one person.

Project participants will frequently present a plan that seems absurd to you. It may in fact be absurd. But perhaps the person who prepared it is simply emphasizing activities that you are not stressing.

Heed others' plans and suggestions.

A common project activity, ordering required materials, illustrates this problem. Sensible choices are to order these materials as early as possible (to be certain they are available when required) or as late as possible (to reduce the possibility of having to change selection or to help your organization's cash flow). It is important to discuss the perceptions of everyone involved in the undertaking.

Or imagine your boss asks you at 9 A.M. to join him or her on a 5 P.M. transcontinental flight to attend an important meeting. You concur and agree to meet at the airport. This arrangement allows you to drive past your house en route to the airport and pack your suitcase. In this simple example, your plan might be as shown in Figure 5-5. For this case, the sequence of activities A, B, and C may not seem to matter. And it does not in terms of your time. But it is desirable to start your secretary's assignment as early as possible so he or she can perform activity D while you do activities B and C. Thus, you should perform activity A first, but you can do C before B (or vice versa) without any schedule delay.

Hazards

There are innumerable hazards in preparing project plans. One follows from the preceding discussion. In an attempt to gain time in the early phases of a project or because you are addicted to your own ideas, you may tend to do much of the planning yourself. You should avoid doing so for the same reason you do not like to be told to carry out somebody else's plan: It is demotivating. In fact, it is important to involve the people who will actually be doing the work so they plan as much of their work as possible. Again, this is the Golden Rule.

Planning their own work can motivate people.

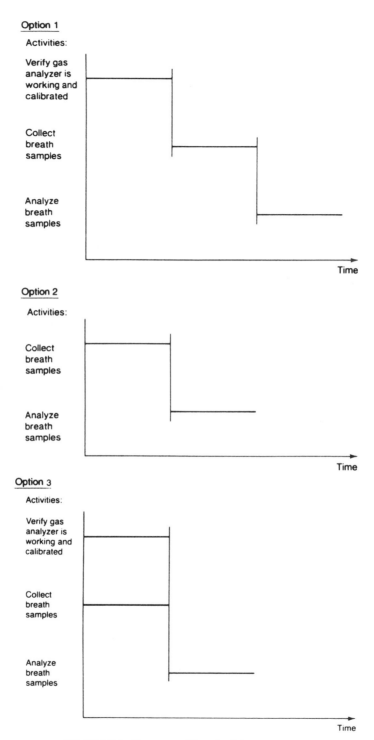

FIGURE 5-4. *Three possible schedule sequences.*

Activities:

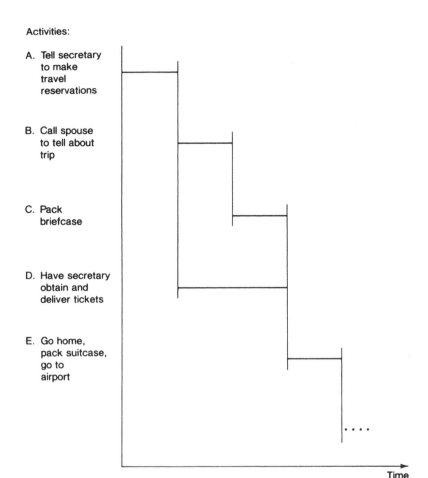

A. Tell secretary
   to make
   travel
   reservations

B. Call spouse
   to tell about
   trip

C. Pack
   briefcase

D. Have secretary
   obtain and
   deliver tickets

E. Go home,
   pack suitcase,
   go to
   airport

Time

FIGURE 5-5. *Option for trip preparation.*

In addition, poor planning frequently occurs. Other than sheer laziness, the basis of almost all poor planning is a misunderstanding of the Triple Constraint point. Taking the time to create plans allows you to identify your perception of the Triple Constraint point and shows if and where it differs from somebody else's.

A good planning technique is for the project manager and the core team to create the project work breakdown structure and identify the task interdependencies using project management software and large-scale group display (for example, either a liquid crystal tablet on a transparency projector or a computer projection system) so that all can see and discuss the plan cooperatively. In practice, this may require several iterations: The first attempt should be aimed at getting all the tasks and their interdependency listed without regard to the actual or expected duration of each task. A second iteration modifies the first by including expected times (or calendar dates) for each task. A third iteration adjusts the second to reflect realistic resource

Task or Activity ╲ Function	Marketing	Engineering	Quality Assurance	
Market Research	X	X		
Product Specification	X	X	X	
Preliminary Design		X		

FIGURE 5-6. *Partial example of a planning matrix, showing involvement of groups in project activities.*

availability. A fourth, and hopefully final, iteration then can adjust everything to come up with a satisfactory and believable compromise. An alternative to using a computer and projection device (or as a preliminary step prior to loading the data into software) is to have the key project team members use a roll of wide paper affixed to a wall, attaching a Post-it note (or some such device) for each task so these are arrayed left to right in time sequence with dependencies drawn on the paper between the tasks. (3M's Planning Roll Kit 551, which provides erasable paper rolls, dry erase pens, an eraser, and Post-it notes, can be conveniently used to assist this group activity.)

Occasionally, a tool commonly called a planning matrix is used. Figure 5-6 provides a partial illustration of such a matrix. It lists activities to be carried out along one side of a piece of paper and designates involved personnel along the perpendicular side. Where these rows and columns intersect at a check mark, the designated personnel are involved in the designated activity. This kind of document may be helpful to some managers, but it is a misnomer to call it a planning matrix rather than an involvement matrix. To put it another way, a planning matrix may be a helpful document, but it is not a plan. The benefits of a planning matrix can be obtained by including a responsibility matrix within computer software (specifically, Microsoft Project), as illustrated in Chapter 13.

Currentness

Once you have decided to plan your project and have issued the plans, people should take them seriously. They can do so only if they know the plans are current.

Therefore, it is very important to know who has copies of them. When you revise plans, be absolutely certain to provide revisions to all people who have copies of previous plans. When you do this conscientiously, everyone involved in your project will know that you take planning seriously.

They will know the plans they have reliably indicate the project intention. You can increase others' assurance by dating all planning documents, and revisions *must* have a revision serial number and date.

"THE PLAN"

In many large projects, there is a book called "the Plan" or "the Project Plan." It may be one or a series of thick notebooks.

Issues Addressed

"The Plan" addresses many topics. It frequently describes what is to be delivered as a result of the project and any specifications for those items. If there will be acceptance tests, it will describe them in detail. In general, the purpose of this kind of plan is to describe what is to be done, by when, and for how much, in other words, to expand upon the Triple Constraint.

Topics Covered

Typically, "the Plan" will cover many of the following topics:

1. Project summary
2. Project requirements
3. Milestones
4. Work breakdown structure
5. Network diagram of the activities with schedule dates
6. Budget for all activities
7. Project management and organization charts
8. Interface definitions, including facility support
9. Logistic support
10. Acceptance plan
11. Standards for property control and security
12. Customer organization contact points, if relevant
13. Nature of project reviews

Activation

Project plans require activation. First, obtain whatever higher level approvals are required, including those of the customer. Second, disseminate "the Plan" to all involved personnel. In very large projects, dissemination may require a chart room in which the walls are covered with charts displaying the plans for and status of various activities, including financial progress and resource allocation. Chart rooms are not required for smaller projects.

> **Spend no more time planning than you would spend correcting problems resulting from having no plan.**

Project plans may vary from fairly simple one-page statements to records with overwhelmingly intricate levels of detail. There is an appropriate level for each project undertaking. There is no magic formula that establishes the right level of detail; in general, never spend more time planning than it would take to correct any problems encountered because planning had not been undertaken. That is, a basic purpose of planning is to avoid problems.

PROJECT PLAN ELEMENTS

Figure 5-7 shows how several project plan elements are related:

1. The first requirement is a complete statement of work (SOW). This may be prepared by the performer, the sponsor, or, most commonly, both together. For a new product project or program, the product requirements determine the performance specification. This may be prepared by an individual in any functional department or by a team within any single department, but it is best prepared by a multifunctional team employing market-based quality function

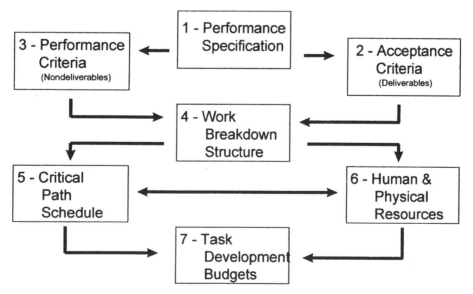

FIGURE 5-7. *Relationships between project plan elements.*

deployment (QFD) or a similar specification-setting process. (See Chapters 2 and 3).

2. Specific, measurable acceptance criteria for all deliverables must be explicitly stated. These are based on the SOW and performance specification and provide an unambiguous target for the development team to aim at.

3. There may also be performance criteria or standards that must be compiled that are not deliverable as such. For instance, your company may wish to comply with ISO 9000; may have environmental, safety, and health (ES&H) requirements to be satisfied; may require that special procurement practices be observed (for example, use standard parts or specific vendors); may presume that special quality standards will be observed; or may presume that corporate styling standards will be maintained.

4. The work breakdown structure is derived from these first three items. The WBS identifies every work package (or activity or task) that must be completed to satisfy the performance specification and other criteria. Most of the WBS will derive directly from the SOW or performance specification itself. However, as one example, the acceptance criteria may stipulate certain test conditions. Thus, additional WBS elements may be required to prepare test facilities or other means to verify completion. Similarly, the performance criteria for nondeliverables may require the inclusion of further WBS elements. (See Chapter 6.)

5. Every element in the WBS must be entered into a critical path network. Milestone, bar (Gantt) chart, and network diagram schedules may be derived from this. (See Chapters 7 and 8.)

6. A resource plan is required to accomplish the WBS. The resource plan includes both human and physical resources. The schedule depends on the resources that will be committed, so these two plan elements interact. For instance, a senior person with relevant experience should be able to carry out a specific task more quickly than an inexperienced junior person. Similarly, if certain resources are overcommitted when the work is to be done, the schedule will be unrealistic and must be adjusted. The resources available for a project depend on priorities and the resources dedicated to other work. Consequently, any project can be delayed by other corporate activities. (See Chapter 10.)

7. Budgets for each task can be created, based on the resources to be dedicated to the task and when that task is scheduled. If appropriate, a benefit-to-cost calculation or a discounted cash flow analysis can then be completed to determine if the project is justified. (See Chapter 9.)

USING COMPUTER SOFTWARE

Project management software of any sort allows you to enter data so that the work breakdown structure, schedule, and budget are all consistent. Specific software packages allow you to display these data in varied formats.

The convenience, economy, and proliferation of computer-based project management software encourages frequent minor plan revisions. In fact, there is a tendency for occasional major overhauls of the plan. This exacerbates the potential for different people working on the project to have different versions of the plan. Obviously, if the plan is maintained in a local area network (LAN) and the database has file lockout, you can be assured that everyone has the same version. However, you cannot be sure that key persons have looked at their portions of the plan recently and noted any changes that affect their work.

TYPICAL PROBLEMS

There are many pervasive problems with planning. Some of them arise because some project team members believe that project management activity per se occupies too much time in general and too much of their time in particular. First, taking enough time to plan is costly. There is an old saying about this: "We don't have enough time to plan now; but we'll have lots of time to fix it up later." In fact, a little inexpensive planning early usually avoids a lot of very costly fixing later. It is axiomatic that software bugs are very expensive to find and remove when a programming project is far advanced. This is analogous to testing engineering breadboards and prototype models—a lot of early testing is much cheaper than trying to remove production problems later. It is difficult to decide how much planning is appropriate, but the inexperienced project manager usually does far too little.

Second, plans are frequently ignored because they are perceived as an irrelevant requirement of management. The solution is obvious: Write meaningful plans you intend to follow and keep current, and be sure everyone understands you have done so.

Third, a separate plan is required for each of the three dimensions of the Triple Constraint. These three plans must be integrated and consistent and must not—as is frequently the case—be prepared by separate groups of specialists. Project management software can help assure consistency.

Fourth, some situations require premature "planning." That is, someone needs a preliminary estimate, typically to establish a budget far before the actual project work will begin. This is especially prevalent where government funding is required. In many cases, the need for the budget estimate precedes the definition of project requirements by one or more years. Thus, in these cases, the budget is established, and, later on, the actual project work cannot be done adequately within the budget constraint. This forces the project manager to seek additional funding or compromise the performance goal.

Fifth, some project managers are unwilling to prepare plans, formal or informal. These people should be reassigned to some other role because a project plan is an absolute requirement of success.

(continued)

Sixth, it is difficult in some cases to plan what work will be required. Team planning is the best solution because more people are likely to recognize all the tasks. However, beware of some people adding tasks that are not required to create work for themselves or their department.

Seventh, it takes time to plan, and many people would rather just get started with the work. This can be overcome by fiat if persuasion is not sufficient.

Eighth, plans that violate the Golden Rule are sometimes made. Consider the following comment from a support group manager:

> As a team member responsible for assembling instrumentation and data acquisition equipment for project use, I face a recurring problem with some project managers. There seems to be an almost universal belief among members of that group that instrumentation issues are of little consequence and that any problems that may occur will be easy to resolve. When I am given a project assignment, I often find that little or no thought has been given to instrumentation availability, cost, or possible lead time.

Early consultation with all people who will be involved and appropriate management can avoid most of these problems.

Ninth, many schedules and budgets are overly optimistic. The solution is to include realistic contingency.

MATERIALS STUDY PROJECT

As project manager for the SUPROMAC materials study project, you have to decide what kind of planning to do. The next four chapters illustrate the work breakdown structure, schedule, and task cost estimates you should create and use. Because it is not a complex project, you can reasonably omit "the Plan" document and a chart room.

However, the research department will be responsible for the theoretical studies that were substituted for testing more materials. Thus, you will arrange to involve them in the detailed planning of their work. (Clearly, there should have been some discussion with them earlier, when this idea arose in negotiations with NERESCO.)

HIGHLIGHTS

- Plans are used for the following purposes: simulate how the project will be carried out; write the proposal; negotiate the contract; coordinate and communicate; increase motivation of participants; control the project; satisfy requirements; avoid problems; and record the choice between options.

- Plans delegate portions of the Triple Constraint to the lowest reporting level.
- Plans help keep projects on course.
- If formulated only to meet someone's requirements for them, plans are virtually useless.
- Everyone involved must receive every plan revision.
- Planning notebooks, a series of planning notebooks, or a chart room may be needed.
- Never spend more time on a plan than would be required to correct problems resulting from a lack of a plan.
- Plans reflect the balance between risk and contingency for both controllable and uncontrollable future events.

6

The Work Breakdown Structure

The goal of a performance dimension plan is to be sure that everything required to satisfy the entire performance specification is done. This chapter deals with planning for the performance dimension of the Triple Constraint. The statement of work is a useful aid, but the principal tool discussed is the work breakdown structure.

STATEMENT OF WORK

The statement of work (SOW) is that portion of the contract that explicitly enumerates what the contracting organization will do for and deliver to the customer. In projects within your organization, the SOW may be contained in a memo or work authorization rather than in a contract, but it should still contain a specific, measurable, and attainable goal. The SOW must always contain a list of all deliverables and should be accompanied by a project schedule and budget to be meaningful. Thus, a plan for the performance dimension of the Triple Constraint is primarily a listing of every activity that must be performed and every result that must be obtained. The SOW frequently contains explicit acceptance criteria and test specifications.

WORK BREAKDOWN STRUCTURE

Purpose

The work breakdown structure (WBS) is a convenient method for dividing a project into small work packages, tasks, or activities. A WBS reduces the likelihood of something dropping through a crack. To put it another way, a WBS is intended to assure that all the required project activities are

Use a WBS to subdivide projects into tasks.

71

FIGURE 6-1. *Work breakdown structure.*

logically identified and related. The production process for a new microprocessor can require hundreds of process steps, some new and specific to that microprocessor's design, which must work together to assure economically adequate yields. In a project to build a new production facility for such a device, first identifying and then executing all the required tasks are crucial to timely completion of the facility.

Figure 6-1 is a WBS, in this case, for a photovoltaic solar power system. There is no magic formula for constructing a WBS. Figure 6-1 shows two levels of detail, but there is no standard number of levels to use. In general, probably at least three or four should be shown, but it might sometimes be appropriate to show five or ten or even more. The breakdown might occur using earlier or later activities, particular organizational involvements, or almost anything that makes reasonable sense. (Note that Figure 6-1 includes the hardware for a photovoltaic solar power system, but the WBS does not list a shipping container, the installation and operation instructions, a warranty document, or a user training manual, some of which might reasonably be a part of such an engineering project.)

In general, it is best to structure the WBS on tangible, deliverable items, both software and hardware. The WBS layout and numbering can be dictated by the customer in the case of some projects awarded by a contract, for instance, those to provide certain weapon systems to the Department of Defense. Figure 6-2 shows two other WBSs, either of which could be used on the breath analysis project of Figure 5-4.

The more work packages you have in your project, the smaller and cheaper each work package becomes. However, the more work packages you have, the more money and time are spent in arranging for these to be properly interfaced with each other and managed. As I discuss more fully in Chapter 17, small WBS tasks with short durations improve the precision of project status monitoring. Conversely, if

FIGURE 6-2. *Two possible WBSs for project illustrated in Figure 5-4.*

you have only one work package, there is no interfacing cost, but the task itself is large and expensive. Therefore, there is a happy midpoint that must be found by experience. In general, you should break your project into work packages sufficiently small so that each is understandable. The more closely each WBS task conforms to prior experience, the more realistic and accurate your plan's schedule and cost estimate become. Another consider- ation in deciding how large a single WBS task should be is whether it will be the responsibility of a senior or junior person and their relevant experience. These considerations and other significant factors are summarized in Table 6-1.

The size of WBS tasks should be chosen carefully.

The WBS defines the work packages and will be tied to attendant schedules and budgets for the work performers. Thus, it is desirable for the lowest level packages to correspond to small work increments and short time periods. It is often helpful to indicate who the task leader is by putting his or her name in the WBS box for the task. It is also possible to insert the WBS number and schedule or budget information. In any event, the WBS can clarify organizational responsibility on a project, as shown in Figure 6-3.

In addition to using the WBS for planning, use task authorization forms, such as in Figure 6-4, to clarify the statement of work for each task. If the task schedule or budget is a constraint, the task authorization form should have a block for this. Note that the task authorization form has a block where the task leader accepts the task, which assures compliance with the Golden Rule discussed in Chapter 5.

The WBS must be tied to time and money plans.

TABLE 6-1   Factors That Affect Task Size

Factor	When to Use Smaller Tasks	When to Use Larger Tasks
Management effort	You can afford to spend more time creating a WBS.	You want to spend less time creating a WBS.
Number of tasks	You want more detail.	You want less detail.
Spending authorization	You want to limit the amount of financial or resource commitment during a specific time interval.	Task managers have previously demonstrated prudent use of money and resources.
Task duration	You want to encourage faster completion of tasks.	You can wait longer to see tasks completed.
Monitoring accuracy	You desire or require greater accuracy.	You can tolerate less accuracy.
Company's prior experience with similar work	Your company has little or no previous experience.	Your company has expertise in specific work of the task(s).
Task manager's skill	You are using an inexperienced person.	You are using an experienced person.

Helpful Hints

In preparing a WBS, do not forget required tasks such as analyses or trade-off studies that must be done but are not specific deliverable items. Also remember to include reports, reviews, and coordination activities. In fact, displaying them on a WBS is a good way to highlight that they are necessary and that resources must be devoted to them.

Fortunately, when a WBS is prepared, it tends to stress hardware integration activities. That is, junctions on the WBS frequently imply a hardware assembly or a test activity that must occur when these things are joined. Thus, a WBS again is useful for identifying an activity to which resources must be devoted.

There is also a slightly different way to create a WBS, which I discuss more fully in Chapter 9. It is a "costed WBS," and the dollar volume attached to each branch of the total project is also included in the WBS boxes (at least at the major headings). This kind of additional information, assuming large amounts of money are associated with large amounts of activity, will direct the manager's attention to those portions of the project that represent the most activity.

The costed WBS is often vital on a project with rigid cost limits. In such a situation, it is often helpful to start with a costed WBS. A trial allocation of the rigid budget is then made to see how well each work package can be performed at that level of funding. Then adjustments are made to the funding allocations to provide the best overall balance of achievable performance within the budget limit.

If you can afford the time, it is desirable to have another person make a WBS for your project, independent of yours, at least down to the third or fourth level. This will take only an hour or so and will highlight any discrepancies or oversights. This approach may suggest a more effective way to organize the required work. You will have to repay the favor on later projects, but that should help your organization by

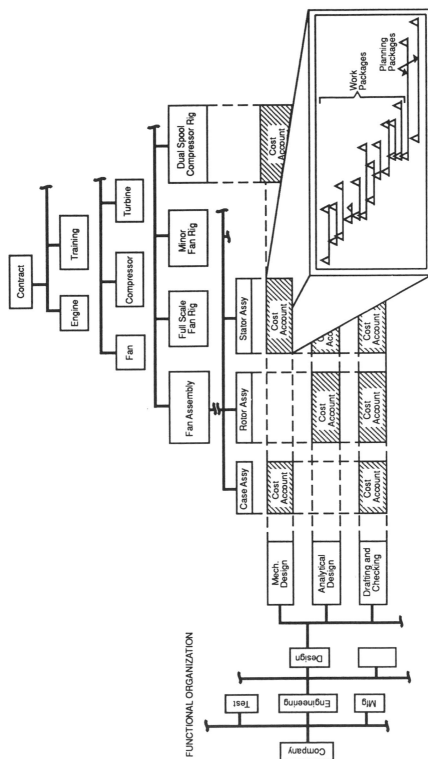

FIGURE 6-3. The integration of a WBS and the organizational structure.

Reproduced from "Cost/Schedule Control Systems Criteria Joint Implementation Guide." Air Force Systems Command Pamphlet AFSCAP/AFLCP 173-5. 1 October 1976.

TASK AUTHORIZATION	PAGE OF
TITLE	

PROJECT NO.	TASK NO.	DATE ISSUED

STATEMENT OF WORK:

APPLICABLE DOCUMENTS:

SCHEDULE

START DATE: COMPLETION DATE:

COST:

ORIGINATED BY:	DATE:	ACCEPTED BY:	DATE:
APPROVED BY:	DATE:	APPROVED BY:	DATE:
APPROVED BY:	DATE:	APPROVED BY:	DATE:

FIGURE 6-4. Task authorization form.

reducing problems on projects. In fact, some organizations require that two or more people independently prepare a WBS for a given customer-sponsored project before it can be approved.

Others can help assure that your WBS is complete.

After the initial WBS has been made, schedule planning can commence. The schedule planning may identify further items to add to the WBS. Although less likely, the same may occur as cost planning is done. The WBS is then revised to include these work packages so that everything on the WBS is finally tied to scheduled work packages and budgets and vice versa.

COMPUTER SOFTWARE

Figure 6-5 illustrates an outline, which is the way several software packages accept WBS information. The ten indented, lettered entries, the tasks that require work,

ID	Task Name
1	START
2	Materials Tasks
3	A - Select Materials
4	B - Obtain Materials
5	Equipment Tasks
6	C - Build
7	D - Debug
8	E - Conduct Experiment
9	F - Document Design
10	Theory Tasks
11	G - Review Literature
12	H - Make Theoretical Study
13	J - Write Theoretical Report
14	Report Task
15	K - Write Final Report
16	DONE

FIGURE 6-5. *A WBS in outline format, created with Microsoft Project.*

are grouped under four summary categories. The first and last entries (start and done) are events, symbolized in this example by the use of all capitals. Events do not require work and are characterized by requiring no time. Every project will have at least the two events shown here, but may also include others (such as customer approvals at certain points).

TYPICAL PROBLEMS

Vagueness in the SOW is a crucial problem in planning the performance dimension. For instance, the SOW may state that "appropriate tests will be performed." Who decides, and when, what is appropriate? The solution is to write a specific and detailed SOW.

(continued)

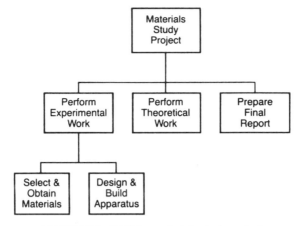

FIGURE 6-6. WBS for materials study project.

TASK AUTHORIZATION	PAGE
	1 OF *1*

TITLE
OBTAIN MATERIALS FOR MATERIALS STUDY PROJECT

PROJECT NO.	TASK NO.	DATE ISSUED
83627	*83627.02*	*3 JAN 01*

STATEMENT OF WORK:
OBTAIN the two materials To be tested in sofficient quantity to perform tests 3 times on each material.

APPLICABLE DOCUMENTS:

See our proposal (copy attached)

SCHEDULE

START DATE: *12 FEB 01*   COMPLETION DATE: *9 MAR 92*

COST: *$9,000.00*

ORIGINATED BY:	DATE:	ACCEPTED BY:	DATE:
APPROVED BY:	DATE:	APPROVED BY:	DATE:
APPROVED BY:	DATE:	APPROVED BY:	DATE:

FIGURE 6-7. Task authorization form for the materials study project.

Another problem is the blind copying of a prior project's WBS for a new project. When this occurs, we have not a WBS, but a waste of everybody's time—the people who prepare the WBS and the people who must read it. A project's WBS should be prepared thoughtfully, not by rote, to increase the odds of project success.

A third problem occurs when you prepare a WBS for any project and fail to revise it as subsequent schedule and cost planning reveal other work packages that are required.

MATERIALS STUDY PROJECT

The materials study project calls for experimental work, theoretical work, and a final report. To do the experimental work, you must test materials in an apparatus. The WBS might be as illustrated in Figure 6-5 and 6-6, and a task authorization for task B might be as illustrated in Figure 6-7.

HIGHLIGHTS

- A work breakdown structure identifies all work packages required on a project.
- A co-worker's independently produced WBS for your project may identify omissions on your WBS.
- The statement of work specifies what the customer will receive and when delivery will occur.

7

Scheduling Tools

This chapter is the first of two that deal with the second dimension of the Triple Constraint. The schedule dimension orders activities so you can identify the logical time relationship between them. In general, there are three approaches to scheduling: bar charts, milestones, and network diagrams. I discuss each in this chapter and stress network diagram usage, especially a format that portrays each time-scaled task with explicit task interdependency linkages. I also illustrate computer project management software that can provide suitable schedule outputs, but caution that its idiosyncratic nomenclature can be confusing.

OVERVIEW OF SCHEDULING METHODS

Table 7-1 provides an overview of the main scheduling methods. Bar charts portray the time schedule of activities or tasks, and milestone charts portray the schedule of selected key events. Network diagrams portray activities, events, or both and explicitly depict their interdependency with predecessors and successors. Time-scaled tasks with explicit task interdependency linkage (TSTETIL) schedules, which I favor, portray both interdependency and the time sequence. In Microsoft Project, TSTETIL is called "Gantt" when that format uses a layout that explicitly connects each predecessor task or event with each successor. You can use the critical path method (CPM) with any of these formats by accenting or emphasizing the tasks (or activities) and events that must be completed on schedule to assure that the entire project is not late.

BAR CHARTS

Bar charts, often called Gantt charts after H. L. Gantt, an industrial engineer who popularized them during World War I, are frequently used for scheduling. Figure 7-1 is a bar chart. The project is divided into five activities with a planned duration of twelve months. When the bar chart was constructed, five open bars were drawn

TABLE 7-1   Comparison of Scheduling Approaches

	Linear Time Scale	
Dependency Picture	No	Yes
No	Lists of tasks or milestones	Bar (Gantt) charts or milestone charts
Yes	Network diagrams— Event-in node (PERT) Activity-in-node (PDM) Activity-on-arrow (ADM) Hybrids with anything in node or on arrow	Time-scaled tasks with explicit task interdependency linkages (TSTETIL)

to represent the planned time span for each activity. The figure also shows project status at the end of the sixth month. The shaded bars represent the forecasted span of the activities as of the end of the sixth month. Activity A was completed early. Activity B is forecast to be finished half a month late. Activity C is forecast to end approximately a month and a half early. The percentage of completion for each activity in process is also illustrated. Activity A has been completed; B is 80 percent complete; and C is 30 percent complete.

Bar charts are simple to construct and easy to understand and change. They show graphically which activities are ahead of or behind schedule.

Offsetting these favorable features are some weaknesses, the most serious of which is that bar charts are essentially useless for *managing* a project (as opposed to providing a

Know the advantages and disadvantages of bar charts.

FIGURE 7-1. *Typical bar chart, illustrating a project with five activities at the six-month review.*

superficial overview of status). Knowing the status of project activities gives no information at all about overall project status because one activity's dependence on another and the entire project's dependence upon any particular activity are not apparent.

In addition, the notion of a percentage completion is difficult and is most commonly associated with the use of bar charts for measuring progress (which I discuss further in Part 4). Does the percentage completion refer to the performance dimension, the schedule dimension, or the cost dimension of the job? Unless an activity is linearly measurable, for instance, drilling a hundred holes in a steel plate, it is impossible to judge what percentage of it is complete. (Even in this simple case, the steel plate may have an internal defect, and the last drill hole might be through that defect, causing the plate to crack, at which point what was 99 percent complete now has to be done all over again.) Therefore, percentage completion becomes highly subjective or is frequently taken merely as the percentage of cost expended compared to total projected cost. In neither situation is percentage completion a useful number. Bar charts are much more useful as an indication of what has happened than as a planning tool to aid the project manager in making things happen properly in the future.

MILESTONES

A milestone schedule notes a few key events, called milestones, on a calendar bar chart. Milestones have been defined in various ways, but they probably are best defined as events clearly verifiable by other people or requiring approval before proceeding further. If milestones are so defined, projects will not have so many that the conclusion of each activity itself becomes a milestone.

> *A schedule that does not show task or activity interdependencies is useless by itself for planning.*

The key to helpful use of milestones is selectivity. If you use only a few key events—perhaps one every three months or so—you will avoid turning milestones into pebbles (sometimes called inchstones) over which people are always stumbling. Some useful milestones might be, for instance, a major design review or a first article test.

When milestones have been defined, for instance, in the customer's request for a proposal or in your proposal document, listing them often helps in preparing your project plan. Having such milestones with attendant schedule and budget measures adds extra emphasis to a few key points of a project. But, in common with bar charts, milestone schedules do not clarify activity or task interdependencies. Thus, they must be used with other tools if they are used at all.

NETWORK DIAGRAMS

> *Networks indicate crucial interdependencies.*

There are many forms of network diagrams, but the Program Evaluation and Review Technique (PERT), the Precedence Diagraming Method (PDM), and the Arrow Diagraming Method (ADM) are the most common. "Network diagram" is a generic term for PERT, PDM, ADM, TBAOA, bubble

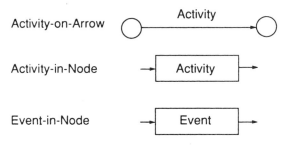

FIGURE 7-2. Three types of network diagrams.

diagrams, and many others. The following list contains some of these and common abbreviations (that I will use):

PERT   Event-in-node (EIN)   Program evaluation and review technique
PDM   Activity-in-node (AIN)   Precedence diagraming method
ADM   Activity-on-arrow (AOA)   Arrow diagraming method
TBAOA   AOA with linear time scale   Time-based AOA

Network diagrams are the recommended approaches to planning the schedule dimension for any project. They identify the precedence conditions and the sequential constraints for each activity.

PERT and ADM emerged in different ways in the late 1950s. PERT is event oriented (that is, the event labels go in the nodes of the diagram) and has typically been used for aerospace and research and development (R&D) projects for which the time for each activity is uncertain. ADM is activity oriented (that is, the activity labels are placed on the arrows) and has been applied to the construction industry, in which there is typically a controllable time for each activity. Figure 7-2 portrays these two forms as well as activity-in-node, most commonly called a precedence diagram. (As I said before, hybrids are common today, primarily due to computer-based project management, software packages.) For instance, an activity-in-node may be zero time, namely, a milestone.

Because of the proliferation of computer-based project management software packages, additional forms of network diagrams will probably continue to appear. Many of these project management software packages provide some form of precedence diagram (without a linear time base), but call it a PERT diagram.

WHAT IS A NETWORK DIAGRAM?

A network diagram is any of several displays that link project activities (or tasks) and events with one another to portray interdependencies. A single activity or event may have interdependencies with predecessor, successor, and parallel activities, or events. Figure 7-3 compares the three principal types: activity-in-node (AIN), activity-on-arrow (AOA, also called task-on-arrow), and event-in-node (EIN). In Micro-

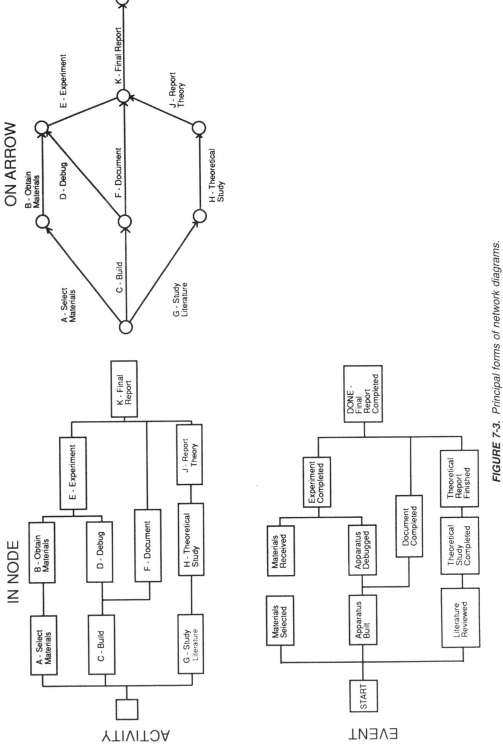

FIGURE 7-3. Principal forms of network diagrams.

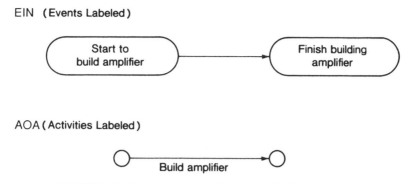

FIGURE 7-4. *Examples of event-in-node and activity-on-arrow.*

soft Project, "PERT" is a hybrid of AIN and EIN and "Gantt" can provide a time-scaled variant of AOA in which arrows are replaced by bars connected with arrows.

One of the few useful things a project manager can do when the project is in some difficulty is to change the allocation of resources dedicated to an activity. Consequently, if you are not already wedded to and successful with the EIN format, adopt one of the other two. As I said before, I prefer time-scaled tasks with explicit task interdependency linkages, which I have abbreviated as TSTETIL.

TSTETIL depicts both interdependency *and* the time schedule. People familiar with bar charts find TSTETIL easier to understand. As you will see, TSTETIL can be constructed to look very much like a bar chart—but it contains and graphically reveals additional essential task interdependency information.

Conventions

Figure 7-4 shows how an amplifier building project would be illustrated in EIN and AOA diagrams. Figure 7-5 provides the symbolic conventions common to both. Activities are always shown as arrows with the start being the tail of the arrow and completion being the barb. Events are shown as circles (or squares, ovals, or any other convenient closed figure). The event number is placed inside the closed figure. Event numbers are occasionally required in computer programs, which are sometimes used to facilitate network information manipulation. In such cases, a higher number activity always follows a lower number activity. In some computer programs, activities are not labeled by their name (that is, "build amplifier" is not called "build amplifier") but rather by the start and end numbers (that is, 5–10 is build amplifier if 5 is the number of the start node and 10 is the number of the finish node). It is also conventional to place early and late date numbers or other information such as slack time within the nodes. Whenever node numbers are used, there should be a legend explaining which number is which. Using these conventions, a network diagram consists of a series of nodes and arrows connected to show the order of activities.

Arrows designate activities or tasks.

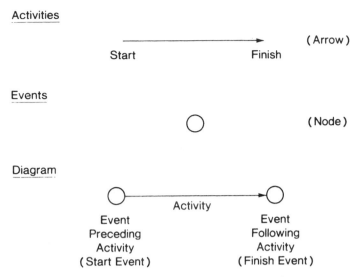

Activities

Events

Diagram

FIGURE 7-5. Basic network diagram conventions.

The upper drawing of Figure 7-6 depicts an AOA schedule plan in which activity R must be complete before activity S can commence and in which activity T must be complete before activity U can commence. The middle drawing shows a schedule plan in which both activities R and T must be complete before either S or U can commence. The bottom drawing introduces the concept of a dummy activity, which is an activity requiring no work, that is, a precedence condition. It thus depicts a plan in which both activities R and T must be complete before activity S can commerce and in which activity T must be complete before activity U can commence. Activity U does not depend on activity R because the dummy arrow points in the other direction.

A dummy activity is a precedence condition.

Network Terms

Figure 7-7 illustrates three terms in network diagram usage. A burst node (node 2) is a node or an event at which two or more activities can be initiated after completion of a preceding activity. A merge node or event (node 5) is one in which two or more activities must be completed prior to initiation of the subsequent activity.

A dummy activity represents a dependency between two activities for which no work is specifically required. Dummies are also used to deal with an ambiguity that arises in some computer-based network diagrams, also illustrated in Figure 7-7. As mentioned, in computer-based network diagram programs, the activity label is sometimes not the activity name but rather the number of the two nodes preceding and following it. Thus, is activity 6–7 task G or task H in the lower left of Figure 7-7. Using a dummy task, one can make activity 6–7 task G and activity 6–8 task

R before S, T before U

R and T before S and U

R and T before S, T before U

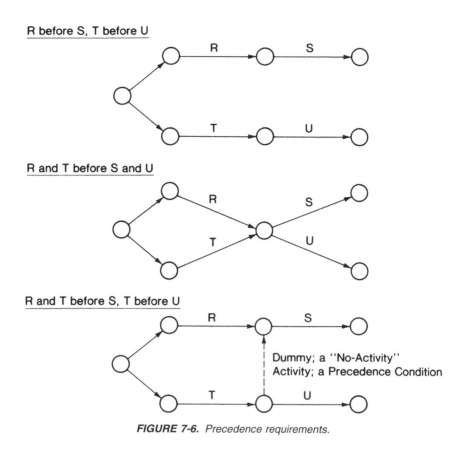

Dummy; a "No-Activity" Activity; a Precedence Condition

FIGURE 7-6. *Precedence requirements.*

H, as shown in the lower right of Figure 7-7. There is thus a dummy activity, 7–8, also required to remove the previous ambiguity.

WHY USE A NETWORK DIAGRAM?

An Illustrative Situation

Consider the project situation illustrated by the bar chart in Figure 7-8. You are the project manager for this project with eight activities or tasks. At the end of four months, you are conducting a project review (denoted by the triangle). Task managers provide status reports (denoted by the shading) showing that tasks B and C are two months late, A and D are one month late, and E is on schedule. The impact of these delays on the entire project's completion is not clear. (For simplicity in this example, assume that task status is precisely measurable—for instance, by counting the number of holes drilled or drawings completed.) Your chief concern now is whether the entire project is late.

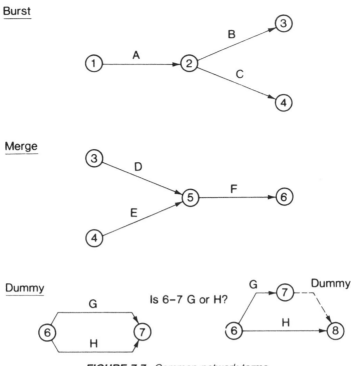

Burst

Merge

Dummy Is 6–7 G or H?

FIGURE 7-7. *Common network terms.*

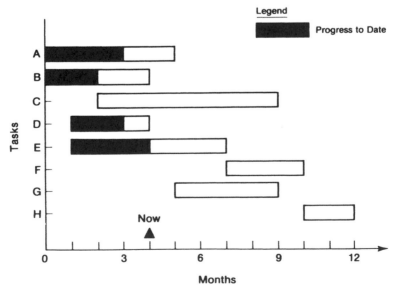

FIGURE 7-8. *Bar chart showing tasks A, B, C, and D to be late at four-month project review.*

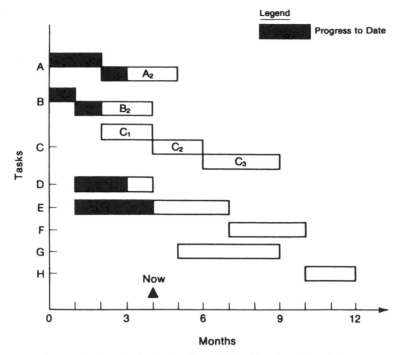

FIGURE 7-9. *Bar chart (from Figure 7-8) with subtask breakdown.*

First you might break tasks into their subtasks or subactivities, as in Figure 7-9. This provides additional information but still does not tell us whether the project will be late.

An Event-in-Node Network Diagram At this point, a network diagram for the project can be examined (Figure 7-10). This contains more information than the bar charts because it shows the interrelationship (precedence) between different tasks. I have labeled each event in its node with the completion of the designated subtask. Thus, at the top of the diagram, you can see that activity D must be completed before activity G.

Networks have more information than bar charts.

Figure 7-10 also shows the problem with an EIN network: The activities per se are not illustrated; that is, there is no arrow uniquely associated with activities G, H, or C₃ This will always be the case where two or more arrows come to a single node, that is, at all merge nodes. This is not a problem for a skilled practitioner, but it does seem to present an unnecessary conceptual difficulty. When the activities are not explicitly shown on a diagram, it is more difficult for the project manager and others to visualize them and their relationship.

A project manager's ability to influence the course of his or her project depends on his or her ability to influence the work on a given task or activity. One of the

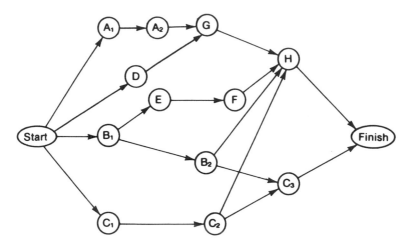

FIGURE 7-10. Event-in-node network diagram for project Figure 7-9. Each node designates the completion of the indicated task or subtask.

few things a project manager can do is change the allocation of resources devoted to a particular activity. Thus, the lack of each activity's explicit visibility in an EIN diagram may be troubling.

An Activity-On-Arrow Network Diagram This EIN problem can be eliminated if we go to an AOA network (Figure 7-11), which shows all activities by labels on the arrows. It clearly indicates the precedences. The requirement of a dummy activity,

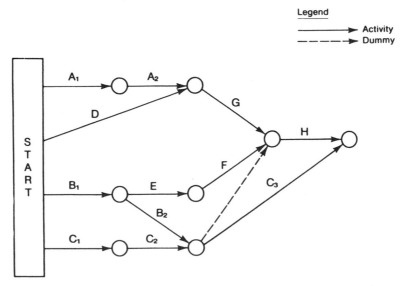

FIGURE 7-11. Activity-on-arrow diagram with each subtask activity and one precedence condition (or dummy).

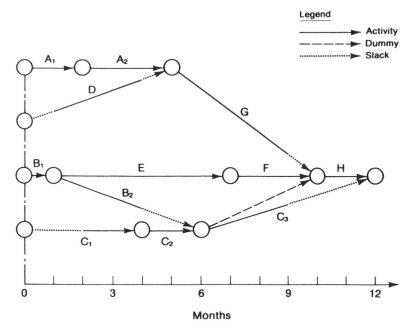

FIGURE 7-12. *TSTETIL diagram, drawn on assumption of task schedules shown in Figure 7-9. (Note that the start node representation is an alternative to that used in Figure 7-11.)*

a "no-activity" activity, is to indicate that the completion of activities B_2 and C_2 (as well as activities F and G) must precede the start of activity H. But in this AOA representation, the merge nodes are not single activity completion events. For instance, the node to which D and A_2 arrows come would not have to be designated the completion of both D and A_2. Frankly, this AOA duality does not trouble me as much as the EIN's lack of activity emphasis and clarity. In fact, the duality may actually be helpful by emphasizing that both D and A_2 must be completed before G can commerce.

Figure 7-12 illustrates the next step in using the AOA diagram: redrawing it to a time scale in which the horizontal projection of each arrow is proportional to the amount of time required for its activity. This is essentially a TSTETIL diagram in which the horizontal distance spanned by an arrow is the time schedule and the predecessors and successors for each task are clearly revealed by junctions. Doing this reveals that one path (B_1, E, F, H) is longer than any other. This is called the critical path. It may also be identified as the path that contains no slack time (amount of time available on a path that is the difference between that required on the critical path and that required on the particular activity path with slack time).

> *The critical path, which indicates the shortest time in which the project can be completed, has no slack.*

Figure 7-12 is drawn with each activity shown starting at the time it was scheduled

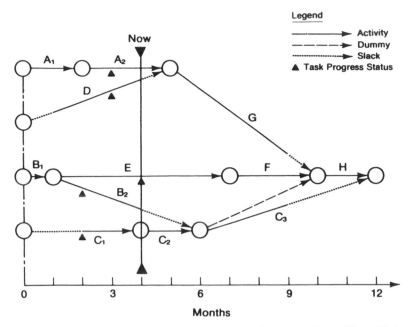

FIGURE 7-13. *Figure 7-12 with task progress status from Figure 7-9 denoted by solid triangles.*

in Figure 7-9. Noting the task progress status and current date, Figure 7-13 begins to provide direct information as to the implications of the delay on activities A, B, C, and D. Figure 7-14 is a redrawn version of Figure 7-13 in which the project manager has taken advantage of the slack time. That is, all late (delayed) activities are drawn to show the work remaining to be done, and subsequent tasks are thus rescheduled in several cases.

Thus, although activities A, B, C, and D are in fact later than planned, the project has not yet suffered any irretrievable schedule slippage. But the project now has two critical paths, whereas it previously had only one. That is, there is no longer any slack in the upper branch (A_2 and G). Because there is still one month of slack on the lower path (task C), perhaps some of the resources allocated to it (or to task B_2) might be redeployed to one of the other critical path activities. It is vastly more difficult to complete a project on time with more than one critical path, and it is unlikely that this project will be completed on schedule, although it is not yet irretrievably lost.

Another Example

Figures 7-15 and 7-16 are an AOA diagram and two versions of bar charts for a house-painting project. The network diagram clearly contains far more information than either of the bar charts—for instance, the dependency of D_1 and D_2 on A_3.

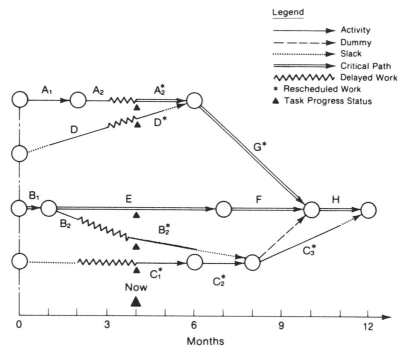

FIGURE 7-14. *TSTETIL redrawn to show delayed work and rescheduled activities, permitting project to be completed in accordance with original twelve-month schedule.*

FIGURE 7-15. *Activity-on-arrow networks for house-painting project, with activity duration in days.*

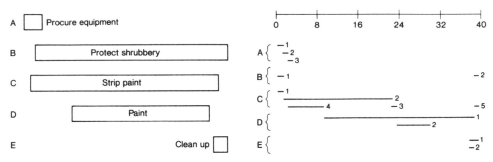

FIGURE 7-16. Bar charts for house-painting project, again revealing lack of task interdependency information.

COMPUTER SOFTWARE

There is no assurance that a project management computer software package will produce TSTETIL. Many produce box displays with both activity and events in the boxes (or nodes), and this may be called "PERT" or "time-based PERT." The latter is not even necessarily linear in time. Figure 7-17 illustrates how Microsoft Project displays a PERT chart (which actually requires two pages of printout, here shown pasted together). The two events (START and DONE) are capitalized; the other ten activities (or tasks) require time. The PERT chart lacks a linear time scale. Figure 7-18 is the Microsoft Project Gantt for the example project, and Figure 7-19 is the same Gantt with the explicit task interdependencies displayed (that is, the TSTETIL version).

> *A critical path network diagram using TSTETIL is the most helpful format for managing a project.*

Another useful feature of project management software is the ability to link tasks or activities in several different ways, as illustrated in Figure 7-20. Each of these time linkages may be zero, plus, or minus time in the most flexible software. I commonly use finish to start with zero time separation, as illustrated in Figure 7-9. However, for a home redecorating project, if task one is shampoo the carpet and task two is move in the furniture, you might wish to have plus two days for the linkage to allow the carpet to dry. No work is required during those two days, but it does affect the overall schedule. Similarly, in the start to start situation, task three might be to perform the test, and task four might be to prepare the test report. Task four can start a few days after task three has started and need not wait for the completion of task three for its initiation.

HELPFUL HINTS

> *Include every element in the WBS in the network diagram.*

One frequently asked question is, "How do I start a network diagram?" One answer is, "With lots of scrap paper." But the best way to start is with the work breakdown structure. From the WBS, you can start the network diagram from either the beginning or the end of the project. There are frequently somewhat obvious large subnets you can quickly

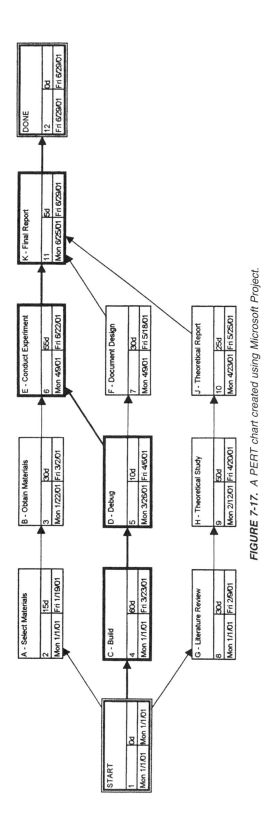

FIGURE 7-17. A PERT chart created using Microsoft Project.

96

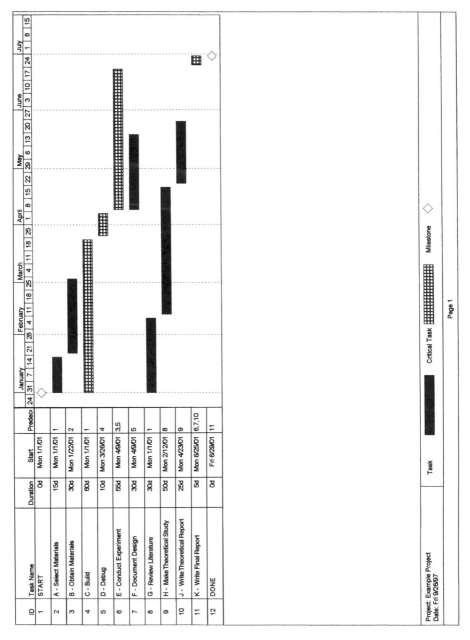

ID	Task Name	Duration	Start	Predec
1	START	0d	Mon 1/1/01	
2	A - Select Materials	15d	Mon 1/1/01	1
3	B - Obtain Materials	30d	Mon 1/22/01	2
4	C - Build	60d	Mon 1/1/01	1
5	D - Debug	10d	Mon 3/26/01	4
6	E - Conduct Experiment	55d	Mon 4/9/01	3,5
7	F - Document Design	30d	Mon 4/9/01	5
8	G - Review Literature	30d	Mon 1/1/01	1
9	H - Make Theoretical Study	50d	Mon 2/12/01	8
10	J - Write Theoretical Report	25d	Mon 4/23/01	9
11	K - Write Final Report	5d	Mon 6/25/01	6,7,10
12	DONE	0d	Fri 6/29/01	11

Project: Example Project
Date: Fri 9/26/97

Task ▅▅▅   Critical Task ▦▦▦   Milestone ◇

Page 1

FIGURE 7-18. A bar chart format layout created using Microsoft Project.

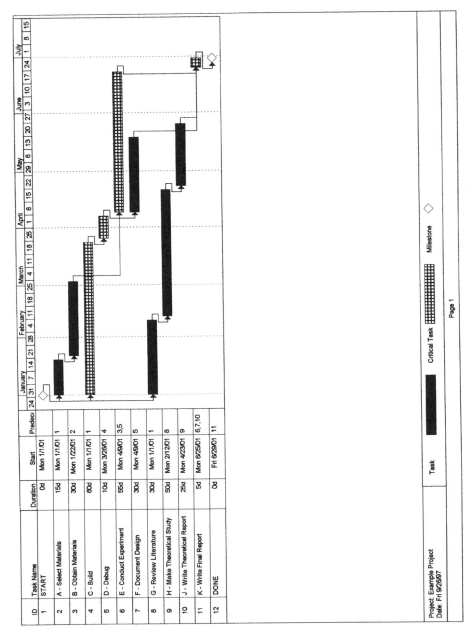

ID	Task Name	Duration	Start	Predeci
1	START	0d	Mon 1/1/01	
2	A - Select Materials	15d	Mon 1/1/01	1
3	B - Obtain Materials	30d	Mon 1/22/01	2
4	C - Build	60d	Mon 1/1/01	1
5	D - Debug	10d	Mon 3/26/01	4
6	E - Conduct Experiment	55d	Mon 4/9/01	3,5
7	F - Document Design	30d	Mon 4/9/01	5
8	G - Review Literature	30d	Mon 1/1/01	1
9	H - Make Theoretical Study	50d	Mon 2/12/01	8
10	J - Write Theoretical Report	25d	Mon 4/23/01	9
11	K - Write Final Report	5d	Mon 6/25/01	6,7,10
12	DONE	0d	Fri 6/29/01	11

Project: Example Project
Date: Fri 9/26/97

Task ▬▬▬ Critical Task ▦▦▦ Milestone ◇

Page 1

FIGURE 7-19. A TSTETIL schedule created using Microsoft Project.

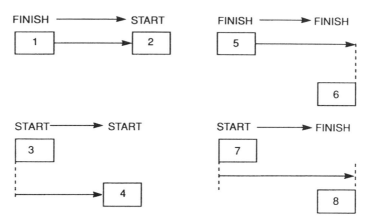

FIGURE 7-20. *Possible task linkages.*

This material originally appeared in a videotape presentation produced by the Association for Media-based Continuing Education for Engineers, Atlanta, Georgia.

put down on a piece of scrap paper. As a general rule, it is probably best to start from each end with scrap paper and sort out the connectedness in the diagram where there are activities in progress simultaneously. Another "paper and pencil" technique is to write the name of each task on a Post-it note, perhaps using different colors for different group's responsibilities. These notes can then be arranged on a large roll of paper in time sequence and these results transferred either to a master schedule or into computer software. As I said in Chapter 6, you may now find some work tasks that should be inserted into the WBS. If so, revise the WBS accordingly.

If you are not using software, you can then transfer the entire diagram to a clean piece of paper. It is probably helpful to do this with a time base and with the presumption that each activity starts at the earliest possible time. For this initial version, I recommend that you get the people who will be responsible for each activity to estimate how long it will take to carry it out on a normal work basis. When you put these time estimates onto the network diagram, it may become apparent that the entire project will take too long. At this point, you can identify particular activities that may be candidates for time compression, that is, tasks you believe can be done faster.

An alternative to time compression is the parallel scheduling of activities, for instance, software and hardware development. There may be increased risk in such a schedule, but that may be the lesser evil. The final step in creating a TSTETIL diagram is to estimate the time of each activity as accurately as possible, using techniques described in Chapter 8.

As shown in Figure 7-21, network diagrams may require crossover of lines. This is to be expected. Although some diagram rearrangement may get rid of crossovers, it may also distort the logical relationship of groups of activities, for instance, all those being carried out by one department being within a general band of the diagram. If activities A, B, C, D, E, F, and G in Figure 7-21 are performed by one section, the upper diagram, which has two crossover intersections, would be prefera-

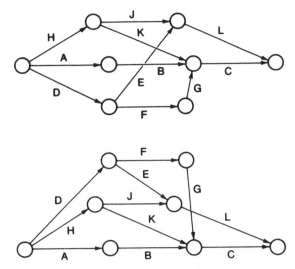

FIGURE 7-21. Activity-on-arrow network diagram with crossovers.

ble to the lower one. Thus, there are cases in which increased use of crossovers will be clearer. Crossovers in some computer project management software can be especially confusing, and there may not be much you can do about it.

The project manager should construct a network diagram of perhaps three dozen activities or up to five dozen if required. Such a diagram can normally be drawn in less than two hours and will fit onto a standard seventeen-by-twenty-two-inch sheet of paper. If some activity in this network is very large, its activity manager can make a network diagram for it. In this way, with a few hand-drawn networks of a few dozen activities each, large projects can be handled without the use of a computer-based network system. Given continuing improvements to and the flexibility of computer project management software, today it may be easier to use software than manual techniques. However, in common with most computer software, project management software improvements and upgrades (for example, the release of Microsoft Project '98 to overcome some limitations in Microsoft Project 4.1) are often accompanied by a bewildering increase in program features and complexity that may be unnecessary for your needs.

Some project (and other) managers resist the use of network diagrams because they consider them complex or because they lack a computer-based project planning network program. This is a serious mistake. It is not the network diagram that is complex: it is the project itself that is complex. In fact, if you can't draw a network diagram for your

If you can't make a network diagram, you can't run the project.

project, that should be a clear danger signal that you do not understand your project. An advantage of using software is that you are forced to confront the logic of your entire project. There is an obvious problem if you cannot load data to get a complete solution, assuming you or a member of your project team is able to use a computer.

I am not opposed to the use of a computer-based planning network program to assist with the mechanics of network usage. In fact, computer programs have great value in determining resource requirements quickly. But a computer-based network diagram program is not required to manage many projects, and the lack of such computer assistance is no excuse for not using a TSTETIL.

Always use a network diagram to plan the schedule dimension, even if you do not show it to your management or customer.

Very large projects, however, will normally require a computer-based system, and many are available. But you usually lose the "hands on" feel you can get by drawing your own network diagram. In general, avoid the computer until you have outlined the project on paper or a planning roll.

TYPICAL PROBLEMS

In many ways, the worst schedule dimension planning problem is to avoid the indicated scheduling problems. For instance, a completed network diagram may show that required materials will not arrive early enough. This conflict is often avoided or dismissed by saying this can be adjusted later. Maybe it can, but that is hoping for luck to save your project schedule. The solution is to admit the problem exists and revise the schedule to overcome it—now, not when there is no longer time to correct the problem and maintain your schedule.

MATERIALS STUDY PROJECT

For the materials study project, in terms of the schedule, assume the following:

- Experiment task
  1. Materials
      A. Select (three weeks)
      B. Obtain (six weeks)
  2. Apparatus
      C. Build (twelve weeks)
      D. Debug (two weeks)
      E. Conduct the experiment (eleven weeks)
      F. Document the design (six weeks)
- Theoretical task
      G. Review the literature (six weeks)
      H. Make a theoretical study (ten weeks)
      J. Write a report of the theoretical results (five weeks)

(continued)

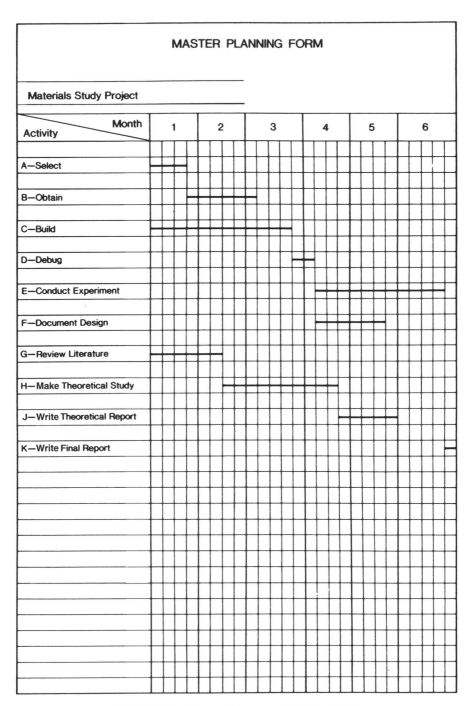

FIGURE 7-22. Bar chart for materials study project.

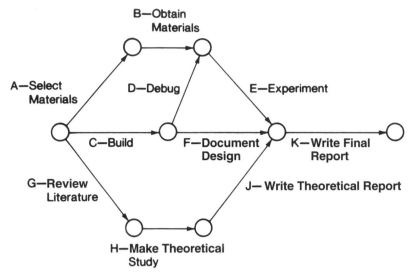

FIGURE 7-23. *Activity-on-arrow network diagram for materials study project.*

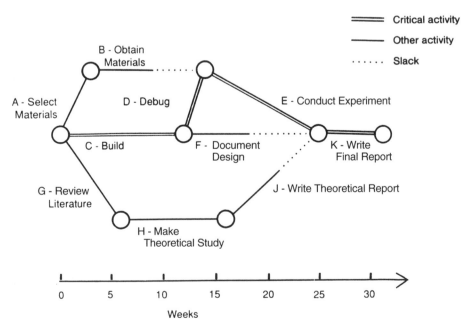

FIGURE 7-24. *TSTETIL diagram for the materials study project.*

> - Final report task
>   K. Write the final report (one week)
>
> A manually created bar chart for this is shown in Figure 7-22. Figures 7-23 and 7-24 are the manually created AOA and TSTETIL diagrams, respectively. Figures 7-17, 7-18, and 7-19 portray the PERT, bar chart, and TSTETIL displays, respectively, as created with Microsoft Project software.

HIGHLIGHTS

- Always use a network diagram, which must include every element in the WBS, to plan the schedule dimension of the Triple Constraint.
- Although easy to make and understand, bar charts and milestones alone are inadequate for schedule planning because they do not show how one activity depends on another.
- Network diagrams show activity interdependencies.
- The most common network diagram forms are PERT, PDM, and ADM.
- Computer software frequently labels its formats uniquely and unconventionally.
- The most effective schedule tool uses a time-scaled task with explicit task interdependency linkages (TSTETIL) format.

8

Time Estimating

This chapter continues the previous chapter's discussion of scheduling techniques. First I cover the important topic of time estimating, which must be done regardless of which scheduling tool is chosen. Next I discuss how a time-scaled task with explicit task interdependency linkages format can clarify the time when individual tasks can or must start, which is crucial for coordinating work.

TYPES OF TIME ESTIMATES

Time estimates are usually inaccurate. Obviously, a schedule for any project requires a knowledge (or estimate) of how long each activity or task will take. Because, by definition, the project has not been done before, there is a necessary inaccuracy in such time estimates. (The only way to guarantee meeting a time estimate is to make it infinitely long, in which case the project won't be authorized.)

After gathering the work breakdown structure and milestones that are necessary to generate a project plan, you need to assess their accuracy. How do you judge if the component estimates are ambitious, reasonable, or overly conservative? Although some projects can be estimated based on similarity to past efforts, many are radically different or require development of new technology or processing methods and are thus hard to estimate accurately. The best you can do is strive to be reasonably accurate in your time estimates. If you estimate a modestly large number of tasks this way, there will be compensating over- and underestimates of time (and cost). The total project estimate will have a smaller percentage error if your team makes estimates for a larger number of tasks, assuming random small over- and underestimate errors for individual tasks. If some tasks on your project are identical or very similar to tasks previously performed, it is easier to estimate these. (As I said in Chapter 6, one of the goals of breaking the project into small work packages is to obtain understandable tasks, which also implies they can be estimated.) There are two time estimating techniques you may wish to use: PERT and pragmatic. Keep in mind, however, that there are two issues in time estimating. The first is to establish the number of labor hours required for a given task (which may depend on the specific skill levels of the personnel). The second is to determine the elapsed time for this activity. You must

Labor hours plus nonlabor costs and task duration must all be estimated.

know the former for cost planning and managing the project, but the latter will determine the overall project schedule. However, if a needed resource is not available when required, that shortage can determine the best schedule.

As an example, consider the production of large-diameter transmission lenses for astronomical use. The molten glass is poured in a mold and then may require many months to harden and cool before the solid blank can be ground and polished. No (or very little) labor is required during this lengthy solidification period. Once the blank is ready for grinding and polishing, a skilled optician is required essentially full time to oversee that work.

PERT Time Estimating

PERT networks originated in projects characterized by uncertain times for activities. This problem was dealt with by requiring three time estimates for each activity:

1. The most probable activity time (T_m)
2. The optimistic activity time, namely, the shortest time that might be achieved 1 percent of the time such an activity was carried out (T_o)
3. The pessimistic activity time, namely, the time that would be exceeded only 1 percent of the time such an activity was carried out (T_p)

As Figure 8-1 shows, this permits calculation of the expected time for the activity (T_e). The basis for this calculation, unproven and unprovable, is the rule applied in PERT networks and is, in fact, a reasonable way to estimate. You can also calculate

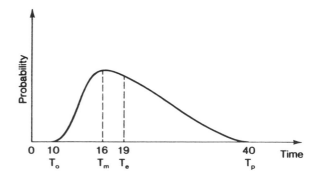

T_m = Most probable time estimate
T_o = Optimistic time estimate
T_p = Pessimistic time estimate
T_e = Expected time for PERT task
$\quad = \dfrac{T_o + 4T_m + T_p}{6}$

$$T_e = \frac{10 + 4 \times 16 + 40}{6} = 19$$

FIGURE 8-1. *PERT time estimating.*

$$T_o = 2$$
$$T_m = 5$$
$$T_p = 14$$

$$T_e = \frac{(1 \times 2) + (4 \times 5) + (1 \times 14)}{6} = 6$$

$$\sigma = \frac{1}{6}(T_p - T_o) = 2$$

FIGURE 8-2. PERT time uncertainty (σ-standard deviation) for a single event.

the uncertainty of that time, which is called the standard deviation (σ). The calculation is illustrated in Figure 8-2.

Figure 8-3 shows how to figure the expected time for a path and standard deviation of the path's expected time. The significance of the calculated standard deviation is the same as with the normal (Gaussian) probability distribution: Two-thirds of the time, the work will be completed within plus or minus one standard deviation of the expected time; 95 percent of the time, it will be completed within two standard deviations; and 99 percent of the time, it will be completed within three standard deviations. This kind of calculation can be important and helpful if there will be a cost penalty for lateness because you can estimate the likelihood of being late.

> **PERT time estimating is useful when time schedule is critical.**

Figure 8-4 illustrates the three time estimates for a particular path containing three activities. Completion of the calculations for this case would show that the expected time is twenty-nine days and the standard deviation is six days. Therefore, the project's completion would be between the twenty-third and thirty-fifth day two-thirds of the time.

Because it requires a lot of effort to make three time estimates and even more time to calculate the expected time and standard deviations, you would normally do this only with a computer-based project schedule plan. (Microsoft Project '98 can perform these PERT time estimate calculations if you provide the three time estimates, but Microsoft Project 4.1 does not include this feature.) Nevertheless, it is important to appreciate the technique and apply it where warranted.

Activity 1   Activity 2   Activity 3

$T_{e,1}$ $T_{e,2}$ $T_{e,3}$

σ_1 σ_2 σ_3

$$T_{e,path} = T_{e,1} + T_{e,2} + T_{e,3}$$
$$\sigma_{path} = \sqrt{\sigma_1^2 + \sigma_2^2 + \sigma_3^2}$$

FIGURE 8-3. PERT expected time and uncertainty for a path.

$$T_o = \quad 4 \quad\quad\quad 1 \quad\quad\quad 2$$

$$T_m = \quad 7 \quad\quad\quad 7 \quad\quad\quad 11$$

$$T_p = \quad 16 \quad\quad\quad 25 \quad\quad\quad 26$$

$$T_e = \left[\frac{4 + (4 \times 7) + 16}{6}\right] + \left[\frac{1 + (4 \times 7) + 25}{6}\right] + \left[\frac{2 + (4 \times 11) + 26}{6}\right]$$

$$= 8 + 9 + 12 = 29$$

$$\sigma = \left[\left(\frac{16 - 4}{6}\right)^2 + \left(\frac{25 - 1}{6}\right)^2 + \left(\frac{26 - 2}{6}\right)^2\right]^{\frac{1}{2}}$$

$$= [36]^{\frac{1}{2}} = 6$$

FIGURE 8-4. Path with three time estimates for each activity.

Pragmatic Time Estimating

I recommend pooled pragmatic judgment for time estimating. The task leader, project manager, and one to three others should discuss the task and arrive at a judgment as to what the schedule should be. The task leader is there because of the Golden Rule. The project manager is there to provide balance with other project time estimates. The others are there to bring expertise and experience to bear.

As a practical matter, the project manager, task leader, and one to three others cannot hold discussions on every task on a large project because there simply is not enough time. In such a large project, this is the goal at which to aim, and a practical compromise is for the project manager to have several deputies to represent him or her in these task estimating meetings.

The goal of such a group estimating meeting is to arrive at a sensible consensus for how long the task will take when it is run in the intended way. If the task leader is a junior person, he or she will not normally be able to complete the task as quickly as a senior person (who may be one of the consulted experts). Sometimes the reverse is true—a junior mechanical engineer may be able to complete a design very quickly using computer-aided design tools, which older engineers have never learned to use.

Base the time estimate for a task on who will do the work and how they will do it.

If a task, as distinct from the project of which it is a part, is identical (or very similar) to a previously completed task, then the experience on the prior task is a good estimating guide. But be certain that there are no meaningful differences that invalidate the relevancy of previous experience.

A logical sequence for estimating a new task is to (1) determine how many days the previous, similar task required and how many personnel worked on it by consulting existing company project records, (2) decide how much more complex the present project is to arrive at a time duration and personnel multiplier, and (3) determine the cost of the new task by multiplying the person days by the appropriate labor rates. This assumes such records exist and underscores the importance of retaining project histories. If no such records are retained, then memory is all that can be used, and memory may differ from one person to another.

What To Do When the Critical Path Is Too Long

In some cases, the logical critical path duration is too long for the sponsor or your organization. You normally discover this when you calculate its duration using the natural (or first trial) time estimates for each activity. You can force the forecast completion date to conform to the schedule by reducing all activities in proportion to the time overrun. But this is usually unrealistic.

The sensible approach is to replan tasks so there is a credible reason to believe they will be completed in times shorter than the first estimate. In general, start with those that are early in the project, and replan enough of these to achieve a satisfactory completion date. (Keep later project tasks at their first time estimate in case you subsequently need to recover from unplanned problems.) In doing this, try to compress tasks that have low compression costs and risks, and be careful not to create another critical path. If this will not save enough time, you will have to entirely replan the project, perhaps running some tasks in parallel. Figure 8-5 shows that other options to shorten the schedule, if that is required, include negotiating for a simpler specification, using more effective resources, and devising a more efficient project plan.

> **When a shorter schedule is required, change the plan, not just the duration of tasks.**

FIGURE 8-5. *More effective resources or a more efficient project plan can shorten the completion schedule for a given specification. See also Figure 2-2.*

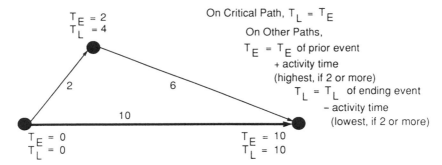

FIGURE 8-6. *Earliest and latest starts.*

This material originally appeared in a videotape presentation produced by the Association for Media-based Continuing Education for Engineers, Atlanta, Georgia.

EARLIEST AND LATEST START AND FINISH TIMES

To introduce this important benefit of network diagrams, consider Figure 8-6. Although the activity durations are shown on the arrows, this is an EIN version, with time emphasis on the nodes. It is conventional to assume that projects start at time equal to zero. The earliest time (T_E) you can emerge from the start node is zero. On the critical path, adding the activity time (in this case, 10), the earliest time you can get to the finish node is 10. Latest and earliest are always the same on nodes for the critical path; hence, $T_L = 10$ at the finish node and $T_L = 0$ at the start node. Off the critical path, the earliest you can reach the upper node is the earliest you can leave the start node plus the activity time on that path (in this case, 2).

Latest times are calculated by working backward. Thus, T_L at the finish node (10) less the activity time (6) determines that the latest time you can leave the upper node (without delaying completion) is $T_L = 4$. The difference between $T_E = 2$ and $T_L = 4$ at this upper node (2) is the slack, or float, on the upper path.

Now consider an AOA network diagram (Figure 8-7). The entire project is always assumed to start at time zero. Thus, the start of each activity that emerges from the start node has zero as its earliest start time (E_S). The earliest finish time (E_F) for each of these initial activities on the critical path is the duration of the activity itself (Figure

> *The difference between earliest and latest times at a node indicates the amount of slack.*

8-8), plus the earliest finish of the predecessor critical path activity. For the entire network, earliest start is the duration of the activity itself (Figure 8-9A). Earliest start and finish times are calculated by proceeding from the start node to the finish node. In Figure 8-9A, activity duration is shown by the number above the middle of the activity arrows. The earliest finish of an activity is equal to the activity duration plus the earliest start. At the merge node, the earliest start of the following activity is the higher of the earliest finishes for the preceding activities. On the

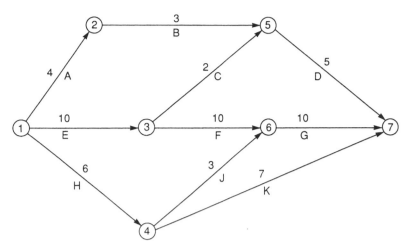

FIGURE 8-7. *An activity-on-arrow diagram showing activity durations (the numbers above the arrows).*

critical path, the earliest finish at the finish node is both the minimum project duration and the latest finish for that activity.

Figure 8-9B shows how to calculate latest finish (L_F) and latest start (L_S) times for each activity. Calculation commences at the finish node and proceeds backward to the start node. In Figure 8-9B, on the critical path, the latest and earliest times are equal. The latest start of an activity equals the latest finish of that activity minus the activity duration. At the burst nodes, the latest finish of the preceding activity is the lower of the latest starts for the following activities.

Table 8-1 shows the kinds of data provided by a typical computer printout for the project illustrated in Figure 8-9. Although less graphic, these data reveal the

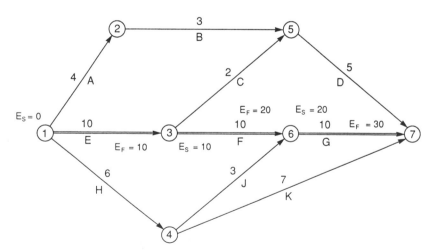

FIGURE 8-8. *The activity-on-arrow of Figure 8-7 with earliest and latest start and finish times on the critical path.*

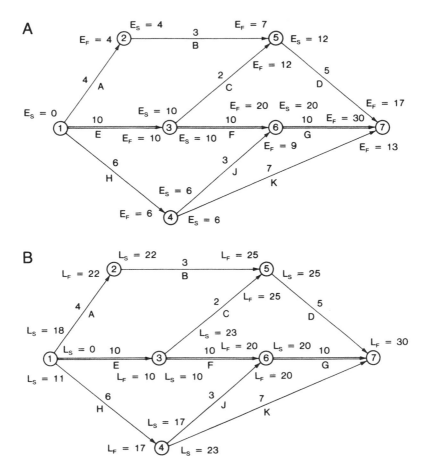

FIGURE 8-9. *(A) Earliest start (E_S) and earliest finish (E_F) calculations. (B) Latest start (L_S) and latest finish (L_F) calculations.*

TABLE 8-1 Typical Data Provided in a Computer Printout for Computer-Based Network Reporting (Use with Figure 8-9)

Event				Start		Finish		
Start	Finish	Description	Duration	E	L	E	L	Slack
1	2	Activity A	4	0	18	4	22	18
1	3	Activity E	10	0	0	10	10	0
1	4	Activity H	6	0	11	6	17	11
2	5	Activity B	3	4	22	7	25	18
3	5	Activity C	2	10	23	12	25	13
3	6	Activity F	10	10	10	20	20	0
4	6	Activity J	3	6	17	9	20	11
4	7	Activity K	7	6	23	13	30	17
5	7	Activity D	5	12	25	17	30	13
6	7	Activity G	10	20	20	30	30	0

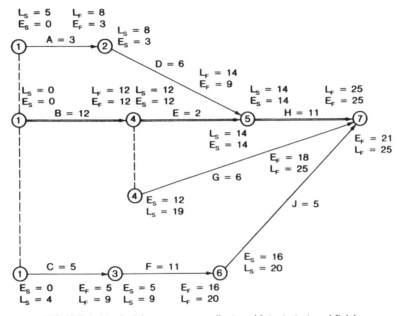

FIGURE 8-10. *Activity-on-arrow earliest and latest start and finish.*

same information. Normally, all the earliest and latest information would be on one diagram. This is illustrated in Figure 8-10. Note that the use of vertical dashed lines, without any dependency arrow indication, permits a node to be drawn in more than one location. This permits spatial separation of activities, thus providing additional open space on the network diagram.

COMPUTER SOFTWARE

Figure 8-11 portrays Figure 8-6 with Microsoft Project. Free slack is the time (or "cushion" following the task) before a delay would make the task critical. Total slack is the slack on the path of which the task is a part.

ID	Task Na	Dura	Pred	Start	Finish	Late Start	Late Finish	Free Slack	Total Slack
1	START	0d		Mon 1/1/01	Mon 1/1/01	Mon 1/1/01	Mon 1/1/01	0d	0d
2	Task 10	10d	1	Mon 1/1/01	Fri 1/12/01	Mon 1/1/01	Fri 1/12/01	0d	0d
3	Task 2	2d	1	Mon 1/1/01	Tue 1/2/01	Wed 1/3/01	Thu 1/4/01	0d	2d
4	Task 6	6d	3	Wed 1/3/01	Wed 1/10/01	Fri 1/5/01	Fri 1/12/01	2d	2d
5	DONE	0d	2,4	Fri 1/12/01	Fri 1/12/01	Fri 1/12/01	Fri 1/12/01	0d	0d

January — February

Task ▨▨▨   Milestone ◇   Critical ▨▨▨

FIGURE 8-11. Example of Figure 8-6 created using Microsoft Project.

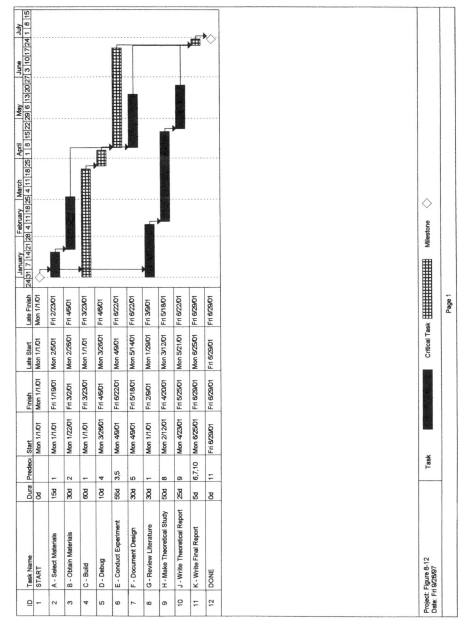

ID	Task Name	Dura	Predec	Start	Finish	Late Start	Late Finish
1	START	0d		Mon 1/1/01	Mon 1/1/01	Mon 1/1/01	Mon 1/1/01
2	A - Select Materials	15d	1	Mon 1/1/01	Fri 1/19/01	Mon 2/5/01	Fri 2/23/01
3	B - Obtain Materials	30d	2	Mon 1/22/01	Fri 3/2/01	Mon 2/26/01	Fri 4/6/01
4	C - Build	60d	1	Mon 1/1/01	Fri 3/23/01	Mon 1/1/01	Fri 3/23/01
5	D - Debug	10d	4	Mon 3/26/01	Fri 4/6/01	Mon 3/26/01	Fri 4/6/01
6	E - Conduct Experiment	55d	3,5	Mon 4/9/01	Fri 6/22/01	Mon 4/9/01	Fri 6/22/01
7	F - Document Design	30d	5	Mon 4/9/01	Fri 5/18/01	Mon 5/14/01	Fri 6/22/01
8	G - Review Literature	30d	1	Mon 1/1/01	Fri 2/9/01	Mon 1/29/01	Fri 3/9/01
9	H - Make Theoretical Study	50d	8	Mon 2/12/01	Fri 4/20/01	Mon 3/12/01	Fri 5/18/01
10	J - Write Theoretical Report	25d	9	Mon 4/23/01	Fri 5/25/01	Mon 5/21/01	Fri 6/22/01
11	K - Write Final Report	5d	6,7,10	Mon 6/25/01	Fri 6/29/01	Mon 6/25/01	Fri 6/29/01
12	DONE	0d	11	Fri 6/29/01	Fri 6/29/01	Fri 6/29/01	Fri 6/29/01

Project: Figure 8-12
Date: Fri 9/26/97

Task ▬▬ Critical Task ▦▦ Milestone ◇

Page 1

FIGURE 8-12. Example project with earliest and latest start and finish dates, created using Microsoft Project.

114

TYPICAL PROBLEMS

There is an interaction of schedule with resources applied. A good designer is usually fast and accurate, and the later fabrication activities may thus be fairly quick. Conversely, a poor or junior draftsperson may be slower and less accurate, and subsequent activities may take longer. The plan reflects what you intend to do, but is not necessarily what you will do. Another aspect of this resource interaction is that having two people on one task is not necessarily as productive as having one person taking twice as long because the two people must spend time comunicating with each other.

Sometimes the project manager or higher management does not like the overall estimated length of the project and wants it reduced. Such a reduction is a problem if time is merely cut out of a task without changing the task's work plan to reflect how this reduction can actually be accomplished.

It is difficult to obtain accurate time estimates for things not done before. You must also note overly ambitious or risky approaches so that backups can be identified and planned in advance. As suggested earlier, getting a few people together, including especially those who will be responsible for the activity, and pooling judgments is the best solution to this problem.

MATERIALS STUDY PROJECT

Figure 8-12 uses Microsoft Project to provide the earliest and latest start and finish dates for the example project. (For space reasons, the free and total slack columns have been omitted.) This TSTETIL format is such that a glance is sufficient to reveal that there is considerable slack on the task A and B path, following task F, and on the task G, H, and K path.

HIGHLIGHTS

- Time estimating is necessarily inaccurate and can best be done pragmatically.
- A network diagram can provide the earliest and latest start and finish dates for each project activity, as well as the free and total slack time for each activity.

9

Planning the Budget

Having a cost plan can help you avoid a situation where actual project cost overruns the estimate or you fail to get the job because you overestimated costs during the proposal and negotiation phase. This chapter tells you how to estimate costs and describes several of the most commonly used cost systems.

COST ESTIMATING

Costs may be stated only in terms of the number of labor hours required, a situation not uncommon in a research group in which a certain number of labor hours have been allocated to a particular project. Cost is more commonly stated in dollars (or yen or marks), however, which entails converting labor hours into dollars. Different hourly rates typically prevail for different seniority levels, and the cost of nonlabor elements (purchases or travel, for instance) is also included. Figure 9-1 shows one way to summarize and total time-phased labor and nonlabor estimates for a task by using a spreadsheet. This figure illustrates the main elements of any successful cost estimating system: Estimate *labor hours* (perhaps by category) and *nonlabor dollars* for each task in each involved department or group. In Figure 9-1, the estimated entries are in larger, italicized numbers. Figure 9-2 is another spreadsheet example. There the series of tasks comprising the project (in this illustration, from preproposal inception to postcontract closure) is listed in the left column.

Cost is, of course, necessary for planning a project, both to sell and manage the job. In general, do not plan costs in detail greater than what you will receive in accounting cost reports. There is no point making cost plans on a daily basis if the organization's cost reports are furnished biweekly or monthly. Cost plans, regardless of how they are arrived at, should typically be summarized in periods corresponding to expense reporting. In counting such things as travel cost or computing hours, however, work with hours or days of travel in estimating and sort these into monthly periods.

Plan costs to the level of detail to which they will be reported to you.

Project: *Materials Study*   Task: *B - Obtain Materials*   Department: *Materials*

LABOR	$/Hour	Hours Each Month						Totals	
		1	2	3	4	5	6	Hours	Dollars
Sr. Prof.	25	8	4	2				14	350
Jr. Prof.	20		40					40	800
Sr. Tech.	15							0	0
Jr. Tech.	10							0	0

DOLLARS	Rate %								
Labor Cost									1150
Overhead	100								1150
Direct Nonlabor		200							200
Subtotal = Prime Costs									2500
G&A	15								375
Subtotal = Total Costs									2875
Profit	10								288
Total = Contract Price									3163

Assumptions:

Prepared By: Date:
Approved By:

FIGURE 9-1. *Typical task cost estimate, using a spreadsheet form.*

Just as with the schedule dimension plans, there are inaccuracies inherent in cost estimates, and these must be expected and tolerated. But tolerating such inaccuracies does not mean encouraging them. The goal is to be as accurate as possible and to recognize that perfection is impossible.

The more effort you put into a cost (or time) estimate, the more accurate it is likely to be. Figure 9-3 indicates the kinds of trends you might expect. In the case of capital construction projects, cost estimates range from about 0.5 to 5 percent of the project's size if you desire accuracy within plus or minus 5 percent. Conversely, the estimates require only about 0.01 to 0.10 percent of the project's size if you need plus or minus 50 percent accuracy.

Techniques

"Forecasting" and "estimating" are frequently used interchangeably to refer to preparing a plan for the cost dimension. Actually, the dictionary definitions of these words are somewhat different. In project management, we are talking about the amount of money (or time) expected to be required to complete a piece of work.

If you were asked to estimate *pi,* you could do so as accurately as you wish because *pi* is a known quantity (3.14159 . . .). If, however, you were asked how long an untelevised football game will take, you would probably reply two and a half hours. You are now estimating a future event's duration

Week # =	1	2	3	4	5	6	7	8	9	10	11	12	13	14	15	16	17	18	19	20
Identify prospects																				
Hours =	4																			
Nonlabor $ =																				
Meet a prospect																				
Hours =		6																		
Nonlabor $ =		50																		
Develop ideas for a project																				
Hours =			12																	
Nonlabor $ =																				
Submit proposal																				
Hours =			6																	
Nonlabor $ =			50																	
Get contract																				
Hours =			2																	
Nonlabor $ =																				
Do the project work																				
Hours =				8	16	16	12	10	8											
Nonlabor $ =					100	100	50	50												
Finish the project																				
Hours =									8	8										
Nonlabor $ =																				
Close the project																				
Hours =											4									
Nonlabor $ =																				
Evaluate the project																				
Hours =											2	2								
Nonlabor $ =																				
Follow up with client																				
Hours =													1		3		1			3
Nonlabor $ =															50					50
Cumulative Totals																				
Labor Hours =	4	10	30	38	54	70	82	92	108	116	122	124	125	125	128	128	129	129	129	132
Labor @ $50/hour =	200	500	1500	1900	2700	3500	4100	4600	5400	5800	6100	6200	6250	6250	6400	6400	6450	6450	6450	6600
Nonlabor $ =	0	50	100	100	200	300	350	400	400	400	400	400	400	400	450	450	450	450	450	500
Dollars =	200	550	1600	2000	2900	3800	4450	5000	5800	6200	6500	6600	6650	6650	6850	6850	6900	6900	6900	7100

FIGURE 9-2. Cumulative labor hour and expense planning with a spreadsheet.

118

FIGURE 9-3. *The expected relationship between the cost to prepare a cost estimate, project size, and the accuracy of the estimate.*

based on similar previous events. You might have made this estimate by looking up the time for the longest and shortest football games ever played and by noting the times of all other football games for which durations were recorded. You would have learned that the vast majority of football games took between two and a quarter and two and three-quarter hours; therefore, two and a half hours is a reasonable estimate.

Actually, the football game you will see will not take two and a half hours. The probability of your estimate being correct is essentially nil. The only way to *guarantee* that actual costs do not exceed your estimate is to make the cost estimate very, very high—which will probably mean there is little likelihood of getting the project authorized. The fact that you are not going to be right means you should become accustomed to being wrong and should not be afraid of it. But it does not imply that you should not try to be accurate. Despite these hazards, the goal in estimating is to have a meaningful plan for your project, one you can use to sell the project proposal to your customer, explain your actions to your boss, and provide enough resources to do the job successfully.

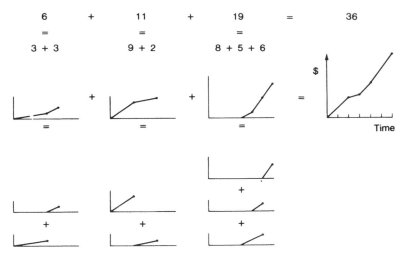

FIGURE 9-4. *A project's cost is the summation of costs for all project tasks.*

Schedule first, estimate second.

There is no point in attempting to estimate a budget for an activity until you have established its duration. In addition, you should understand the preceding and following activities in order to define better the activity you are estimating. Such understanding may clarify that a following activity is farther downstream than it first appears. If so, the activity you are estimating probably is longer, and therefore costs more, than you first thought.

Estimate the cost of each task.

You do the estimating by breaking the project into tasks and activities, using theWBS and network diagrams. The budget of any large activity is the sum of the smaller tasks that compose it, as shown in Figure 9-4. In general, use as much detail as possible. Every task in the WBS should probably have an individual task estimate (such as Figure 9-1) prepared by the responsible task manager.

Shortcuts

There is a variety of means to prepare cost estimates. Using as much detail as possible is commonly called the "bottom up" method. The major project is divided into work packages small enough to allow accurate estimation. The project estimate is the sum of the estimates for all the individual work packages.

There are shortcuts to estimating some of the small work packages. You can use similarities to and differences from other tasks to shortcut a complete level of detail for a second task. Or you can use ratios or standards to relate one small task to another.

Whenever you use the bottom up technique, judge it against a "top down" estimate. The top down estimate is done first, quickly and judgmentally, and then it is set aside. For instance, assume the bottom up estimate comes to $10 million. Your top down estimate, which you now retrieve, is $5 million. Go back and look at each individual work package in the bottom up estimate to find out where the excess costs arise. Examine each package to discover to what extent there has been an incorrect assumption as to the amount of work called for. Or your top down estimate may

To make a project cost estimate, add detailed task estimates for each work department and adjust them if the overall summation seems unreasonable.

indicate that the total job should cost $20 million. Explore the details to find out what has been overlooked or what unwarranted simplifying assumptions were made. The role of the top down estimate, which is obviously not accurate, is to provide a point of view from which to scrutinize the bottom up estimate.

Parametric Cost Estimating

To introduce this technique, consider the situation in which you need to obtain some software to support your project.

Figure 9-5 illustrates a relationship between project cost and a computer program's lines of code that we might discover if we examined a large number of previous computer software projects. The shaded area in the figure would surround a cloud of points, each of which represents a particular project outcome.

If we examine these projects in more detail, we could ask how many separate modules were involved. Then the data points could be separated, and we might find a trend as in Figure 9-6, where the lines are the centroids of smaller clouds of data points.

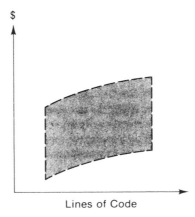

$

Lines of Code

FIGURE 9-5. *Relationship between project cost and lines of program code for a large number of computer software projects.*

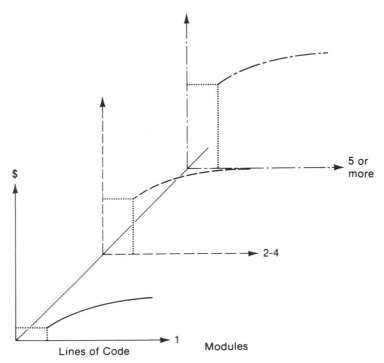

FIGURE 9-6. *Trends of Figure 9-5 data when the number of program modules is considered.*

This is the underlying notion of parametric cost estimating. Historical data for many factors are used:

Project magnitude—How big is it?

Program application—What is its character?

Level of new design—How much new work is needed?

Resources—Who will do the work?

Utilization—What are the hardware constraints?

Customer specifications and reliability requirements—Where and how are these used?

Development environment—What complicating factors exist?

Then, if your project falls within the range of these historical data, you merely insert your project's parameters into the multidimensional regression model to estimate your project's costs.

Parametric model estimates are only as accurate as the input data used.

There are many such parametric cost estimating models. The estimate you get from using such a parametric model is only an estimate and will be inaccurate if your input data (for instance, the resources) are poor estimates.

Cautions

Cost estimates are frequently made prematurely, before the work performance and time schedule are fully understood and defined. Such cost estimates must then be redone or adjusted when the performance and schedule are set. It is clearly more efficient to do the cost estimating after the other two dimensions are defined. However, your project definition may start with a predetermined fixed cost as the highest priority of the Triple Constraint. In that case, you must fit the best work accomplishment and schedule within that cost limitation. A good approach is to use a costed WBS, in which each task box is assigned a trial portion of the specified budget. These arbitrary initial task budgets are then adjusted iteratively as you develop planning detail.

There is a danger signal to which project managers must be alert when receiving cost estimates: the person-month dimension to describe the labor requirement. This is a danger signal because it frequently indicates a snap judgment on the part of the estimator (and not because it is an inappropriate measure, although hours seem preferable). The person who can do the task in two months may not be available when he or she is needed. Or a task that can be done by one person in three months (three person-months) may require four people if it must be completed in one month (four person-months). (So-called "effort-driven scheduling" is a new feature in Microsoft Project '98. If you use that software, be careful because doubling the labor on a task will automatically halve the task time.)

A pertinent issue today is how to plan for inflation. Such planning can be done only with great difficulty and caution. Unfortunately, there are no guarantees in dealing with the future, so some method of coping must be adopted. Make the best estimate of labor hours, regardless of when the activity will occur. Then estimate the rates for these labor hours and the dollar amounts for nonlabor ingredients in current dollars. You can then apply inflation factors for future years to these numbers by consulting with your organization's financial planners.

On lengthy projects, have an inflation hedge.

Another issue that you have to think about is improving productivity, which results from the proliferation of new tools such as computer-aided design and computer-aided engineering. Similar tools exist for software, many forms of white-collar office work, and so on. In many situations, these tools will shorten the time and lower the cost to carry out a given task or activity. However, before these gains can be realized, an investment is required to acquire them (which typically requires a capital expense) and train people to use them. Also, the learning (or experience) curve may apply, in which case future costs may be lower than current ones.

PROJECT COST SYSTEM

To use project cost reports, you must have a project cost accounting system, which is a means to accumulate costs by project and project activity or task detail. If your company does not have a project cost accounting system, you can *plan* your project costs with, for example, project management or spreadsheet software, but you will not have any practical way to *monitor* the actual costs. The following are the elements of such a system:

- Labor
  People in your own department
  Other people in company
- Overhead burden
- Nonlabor
  Purchases
  Subcontracts
  Travel
  Computer charges
- General and administrative burden

There are many variations in detail for project cost systems, some of which are shown in Table 9-1. I am making the simplifying assumption that the organization has three projects (A, B, and C), which all start at the beginning and will end at the close of the fiscal year. It does not matter whether the work is for the company or an outside client or whether it is to be paid for by the sale of goods or contract billing. The point is that there has to be some way of allocating the cost of these three projects to different customers or product lines. The table illustrates four methods of allocating these costs to the three projects.

In method 1, the direct labor and direct nonlabor are allocated to the project, and these are summed to provide a direct total. All the burden and overhead accounts are then lumped and apportioned to each project in proportion to the direct total expenses. In method 1, these are equal, and the billings to each of the three projects would be equal.

In method 2, the direct labor and direct nonlabor are treated as before, but the overhead portion is allocated to each project in accordance with the amount of direct labor it requires. Nevertheless, the general and administrative (G & A) expenses are allocated to the projects in accordance with the direct total, as in method 1. In this case, the billings to the projects are not equal. Project A is more than B, which is more than C.

Method 3 and its variants are the most common project cost systems.

In method 3, the overhead is treated as in method 2, and the direct nonlabor is treated as in both methods 1 and 2. But in this case, all these items and direct labor are joined to come up with a prime cost, and the G & A burden is allocated in proportion to that. In this case, we arrive at a still higher amount of billing for project A.

Method 4 is one of the many common variations of method 3. Direct labor and overhead are treated as in method 3, but purchases are subject to a material handling charge. (In method 3, this material handling charge is included in G & A; in method 4, it is pulled out of G & A and assigned to the

TABLE 9-1 Four Common Project Cost Systems, Illustrated for Three Projects ($000 omitted)

	Project A	Project B	Project C	Indirect		Total
Direct labor	50	30	10		} = 150 { Labor / Purchases / Travel	
Direct nonlabor	0	20	40			} = 270
Overhead on direct labor				90 } = 120 { Fringe benefits / Indirect time of direct labor personnel / Indirect labor personnel / Facility costs / General supplies / Publications		
General and administrative				30		
Burden						
Method 1						
Direct labor	50	30	10			
Direct nonlabor	0	20	40			
Direct total	50	50	50	= 150		
Burdens on direct total	40	40	40	= 120		
Total costs	90	90	90	270		
Method 2						
Direct labor	50	30	10			
Direct nonlabor	0	20	40			
Direct total	50	50	50	= 150		
Overhead on direct labor	50	30	10	= 90		
General and administrative				} = 120		
Burden on direct total	10	10	10	= 30		
Total costs	110	90	70	= 270		
Method 3						
Direct labor	50	30	10			
Overhead on direct labor	50	30	10	= 90		
Direct nonlabor	0	20	40			
"Prime costs"	100	80	60	} = 120		
General and administrative						
Burden on "prime" costs	12.5	10	7.5	= 30		
Total costs	112.5	90	67.5	= 270		
Method 4						
Direct labor	50	30	10			
Overhead on direct labor	50	30	10	= 90		
Direct nonlabor—purchases	0	20	20			
Material handling burden	0	5	5	= 10 } = 120		
Direct nonlabor—other	0	0	20			
"Prime costs"	100	85	65	} = 30		
General and administrative						
Burden on "prime" costs	8	6.8	5.2	= 20		
Total costs	108	91.8	70.8	= 270		

projects in proportion to their required purchases.) But other direct nonlabor, in this case, travel, is not allocated a handling fee, as illustrated in project C. This results in a still different prime cost. Finally, the G & A expenses are again distributed, and a still different billing arrangement is arrived at.

All these methods, and others as well, are used. The project manager must understand his or her company's method in order to know when to use subcontract help and when to use in-house, direct labor.

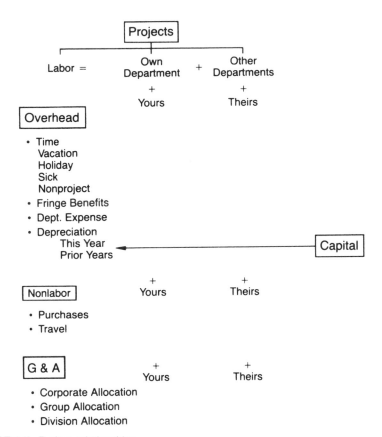

FIGURE 9-7. Budget relationships.

Source: "Basic Management Skills for Engineers and Scientists." Lecture notes by Milton D. Rosenau, Jr. Copyright © 1982 by the University of Southern California. Reprinted by permission.

It is also important to understand any subcontractor's cost accounting system. If you are placing a labor intensive contract with a subcontractor, you should not use a subcontractor who practices method 3 as opposed to method 1.

In addition to understanding project budgets, you should understand the relationship between project budgets and administrative budgets, as illustrated in Figure 9-7. The administrative budgets are overhead, general and administrative, and capital. As we saw in Table 9-1, overhead and G & A may be combined into a single burden pool. Capital expenditures enter into the overhead (or G & A) budgets by requiring inclusion of depreciation. Thus, the direct cost of a project, its own direct labor and nonlabor expenses, is not really a measure of its cost to the organization. The project must, in common with all other organizational activities, carry burdens that depend on other organizational activities and budgets.

COMPUTER SOFTWARE

One of the most attractive features of project management software is that the cost plan automatically drops out of the task detail that you have entered. That is, at the same time you enter schedule information and task linkages, you may enter the resources that will be required, both labor and nonlabor. You may enter labor resources in hours, but you must also provide hourly labor rates if you wish to obtain a monetary cost estimate. Nonlabor estimates, such as purchases and subcontracts, are almost always made in monetary terms, so these will be reported in that fashion. Obviously, this is a classic garbage in, garbage out situation. The soundness of the resulting project cost estimate will be set by the quality of the individual task estimates.

Each software package seems to require these data in a different format, but this is not a problem if you master a single package. In general, you will have to enter all possible resources and their rates at some point if you wish to use this software to plan project costs. Figures 9-8 and 9-9 show a resource cost form and cost report format from Microsoft Project 4.1. (This version lacks a convenient way to spread costs throughout a task unevenly, as illustrated, for instance, in Figure 9-1 for the senior professional. This shortcoming has been remedied in Microsoft Project '98, which offers multiple options for uneven time allocation. An alternative "work-around" is to split task B, as shown in Figure 9-1, into three subtasks: B-1 during month 1, using eight hours of the senior professional and $200 of materials; B-2 during month 2, using four hours of the senior professional and forty hours of the junior professional; and B-3 during month 3, using two hours of the senior professional. This work-around may also be helpful if the options for uneven work rate do not match your actual plan.)

TYPICAL PROBLEMS

There are three important problems in planning the cost dimension of the Triple Constraint. First, many project groups or project managers have a deplorable tendency to make cost estimates for support group work. This forecloses the possibility of benefiting from support group expertise and violates the Golden Rule. This is easily solved by requiring every department to approve the estimate for the work it will do.

A second problem is dealing with inflated estimates by support groups. Here the project manager can first try discussion and negotiation. If that does not produce a satisfactory agreement, the project manager could alter the nature of the requested support work. Two other possible solutions are to subcontract the support work to another company or appeal to higher management.

(continued)

Example Project Cost Est

ID	Resource Name	Initials	Group	Max. Units	Std. Rate	Ovt. Rate	Cost/Use	Accrue At	Base Calendar	Code
1	Engineer	E		1	$20.00/h	$0.00/h	$0.00	Prorated	Standard	
2	Sr. Engineer	S		1	$25.00/h	$0.00/h	$0.00	Prorated	Standard	
3	Designer	D		1	$15.00/h	$0.00/h	$0.00	Prorated	Standard	
4	Researcher	R		1	$25.00/h	$0.00/h	$0.00	Prorated	Standard	

FIGURE 9-8. Resource cost input form for Microsoft Project 4.1.

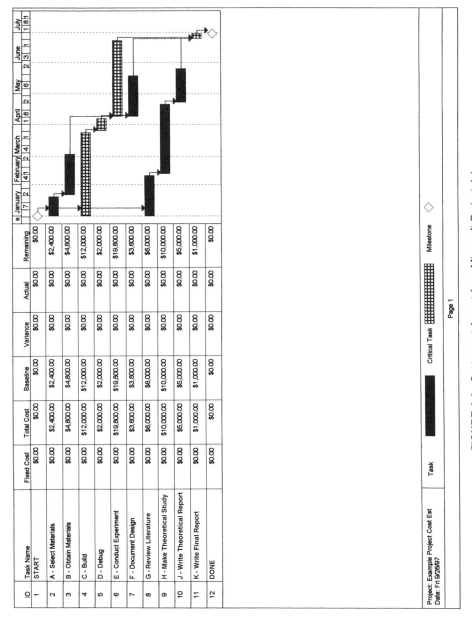

ID	Task Name	Fixed Cost	Total Cost	Baseline	Variance	Actual	Remaining
1	START	$0.00	$0.00	$0.00	$0.00	$0.00	$0.00
2	A - Select Materials	$0.00	$2,400.00	$2,400.00	$0.00	$0.00	$2,400.00
3	B - Obtain Materials	$0.00	$4,800.00	$4,800.00	$0.00	$0.00	$4,800.00
4	C - Build	$0.00	$12,000.00	$12,000.00	$0.00	$0.00	$12,000.00
5	D - Debug	$0.00	$2,000.00	$2,000.00	$0.00	$0.00	$2,000.00
6	E - Conduct Experiment	$0.00	$19,800.00	$19,800.00	$0.00	$0.00	$19,800.00
7	F - Document Design	$0.00	$3,600.00	$3,600.00	$0.00	$0.00	$3,600.00
8	G - Review Literature	$0.00	$6,000.00	$6,000.00	$0.00	$0.00	$6,000.00
9	H - Make Theoretical Study	$0.00	$10,000.00	$10,000.00	$0.00	$0.00	$10,000.00
10	J - Write Theoretical Report	$0.00	$5,000.00	$5,000.00	$0.00	$0.00	$5,000.00
11	K - Write Final Report	$0.00	$1,000.00	$1,000.00	$0.00	$0.00	$1,000.00
12	DONE	$0.00	$0.00	$0.00	$0.00	$0.00	$0.00

Project: Example Project Cost Est
Date: Fri 9/26/97

Task   Critical Task   Milestone ◇

Page 1

FIGURE 9-9. Cost report format from Microsoft Project 4.1.

Higher management, if they decide to "buy in," often cause a third problem. If you are convinced that buying in is disaster, you can request that someone else assume project management. Or you can record your objections in a memo and try to accomplish the promised work within the budget. Finally, you can undertake the job and work actively to sell your customer on changes of scope that provide an opportunity for more funding.

MATERIALS STUDY PROJECT

Each department should estimate its work on each task.

The materials study project consists of ten tasks divided among five departments. Having identified the tasks by the work breakdown structure (Figure 6-5) and the involved departments by reference to the logical work elements and the organizational arrangements, it is then possible to identify which tasks have to be estimated as to cost and schedule by which departments. Figure 9-10 illustrates ways five departments might be involved in the ten tasks and how computers can summarize them.

(continued)

TASK \ SECTION	PROJECT (MECH.)	CHEM.	ELEC.	RESEARCH	TECH. SUPPORT	ALL
A	E	E				S
B	E	E				S
C	E		E		E	S
D	E		E		E	S
E	E	E	E		E	S
F	E				E	S
G	E			E		S
H	E			E		S
J				E		I
K	E	E	E		E	S
ALL	S	S	S	S	S	S

E = ESTIMATE MADE BY INDICATED DEPARTMENT FOR INDICATED TASK

S = SUM OF COLUMN OR ROW PREPARED BY COMPUTER OR ADMINISTRATIVE AIDE

I = SUM OF COLUMN OR ROW IS IDENTICAL TO ONLY ENTRY IN COLUMN OR ROW

S = SUM OF ENTIRE PROJECT, PREPARED BY COMPUTER OR ADMINISTRATIVE AIDE

FIGURE 9-10. *Estimation of the materials study project is accomplished by making twenty-five individual estimates.*

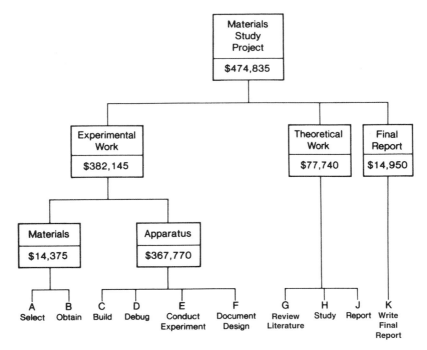

FIGURE 9-11. Costed WBS for materials study project.

> The computer can be used to simplify planning by quickly summarizing large amounts of numerical data. For very small projects, this can easily be done by hand or with a spreadsheet and is not sufficient justification for using a computer.
> Once all the cost estimates are summed by task, it is possible to complete a costed WBS (Figure 9-11) should this be desired.

HIGHLIGHTS

- Cost estimates are usually made in dollars.
- Cost estimates can be made top down or bottom up, but a better method is to do both, in that order.
- A parametric cost estimate may be useful and simple, especially if your project is reasonably similar to others for which historical data are available.
- The elements of a project cost accounting system, a means to tally costs by project and project task, are labor, overhead burden, nonlabor, and general and administrative burden.

10

The Impact of
Limited Resources

This chapter deals with the impact of finite resources on project plans. This topic typically involves two or three dimensions of the Triple Constraint. First I discuss resource allocation and how to resolve resource constraints. Then I present techniques that allow analysis of schedule and budget trade-offs.

RESOURCES

Resources are either people or things. Human resources may include everyone in a particular organizational unit or those with a specific skill (typing, computer programming, senior optical design, analytic chemistry, or journeyman electrician, for example). Things include any kind of equipment, for instance, lathe availability, computer time, or pilot plant time, as well as floor space to house the equipment and people. Money may also be considered a nonhuman resource.

Allocation

There are three reasons to consider resource allocation in a project management environment. First, forecasted use of some key resource (for instance, circuit designers) may indicate there will be surplus personnel at some future period. This information

Surplus resources waste money and talent, so resources are normally overcommitted.

tion should warn the appropriate managers either to obtain new business to utilize the surplus talent or to plan to reassign the involved personnel. However, many resources are not very flexible and cannot easily be reassigned to other useful duties. For instance, not many highly trained specialists are flexible generalists; so an available chemist may not have useful skills for your project that requires an electrical engineer. Because no company can afford to indefinitely maintain surplus resources anticipating the authorization of a future project, the normal situation is that resources are overcommitted. First one resource is overloaded, then another, and so on. Sometimes the overload is only a small amount, sometimes substantial. Sometimes the overload is for only a short period, sometimes for a long period. Failure to allow contingency for this normal overload situation is one reason many projects are delayed.

Resource capacity is limited, and multiple projects may require the same resource at the same time, resulting in a conflict that must be resolved.

Another reason for resource allocation is to avoid inherent inconsistencies, for instance, using a particular resource (Jane Draftsperson, for example) on two tasks at the same time. Preparing a network diagram to a time base emphasizes resource allocation and reveals latent conflicts.

A network diagram can show what resources are required and when, which may reveal that more of some resources will be needed than will be available at some time. When you discover this, you must adjust the network diagram to shift the overloaded resource requirement to some other time. If you fail to do this, slippage will occur. Figure 10-1 illustrates resource allocation. In this case, the resource is the personnel headcount. Tasks A and B, each of eight weeks duration, require three and five personnel, respectively. Tasks C, D, and E are not on the critical path, and examination of the earliest and latest times for them shows they can be commenced immediately or as late as the eleventh week. If the company performing this project employs only six people, task D would

FIGURE 10-1. *Resource allocation.*

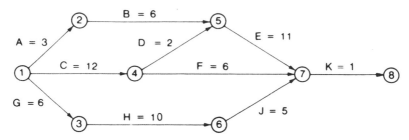

FIGURE 10-2. *Activity-on-arrow diagram.*

have to start early enough to be completed before the end of the eighth week, when task B is scheduled to start. If task D starts later than the start of the sixth week, some (or all) of task D will be scheduled in parallel with task B, requiring seven personnel.

A third use of this kind of analysis occurs in a large company. Imagine that tasks C, D, and E are performed by a particular support department, for instance, the design and drafting section. If the design and drafting section was provided with resource allocation information for all projects, as shown in Figure 10-1, they could identify the earliest and latest dates at which the support, in this case, C, D, and E, would have to be applied. Doing the same for all projects would allow the support group to even out its workload and to identify in each case the impact of any slippage.

Removing Resource Constraints

Consider the AOA network diagram in Figure 10-2. After planning the work, you arrive at a summary of labor skills required for this project (Table 10-1). Suppose that you have only nine junior engineers. What are your options? What might you do? What are the risks?

TABLE 10-1   Resource Allocation, Using Network of Figure 10-2

		Required Resources					
Task	Planned Duration	Senior Eng'r	Junior Eng'r	Design	Mech	Elec	Inspect
A	3	2	4	4			
B	6		2			3	
C	12*	3	4	4			
D	2*		2		5		
E	11*	2	5		2	2	
F	6		2	7			
G	6	4	1				
H	10	1	3				
J	5		2				
K	1*	2	2				3

* = Task is on critical path.

Task	Junior Engineer Required Each Week (Assuming Earliest Start on Each Task)
A	4 4 4
B	2 2 2 2 2 2
C	4 4 4 4 4 4 4 4 4 4 4 4
D	2 2
E	5 5 5 5 5 5 5 5 5 5 5
F	2 2 2 2 2 2
G	1 1 1 1 1 1
H	3 3 3 3 3 3 3 3 3 3
J	2 2 2 2 2
K	2

FIGURE 10-3. *Junior engineer resource requirement on first project network diagram of Figure 10-2.*

Figure 10-3 graphically summarizes the junior engineer staffing each week. There are two weeks when ten junior engineers are required, but the constraint is that only nine are available. Some of your options are to use overtime, use senior engineers to do the work, or delay task F (which has slack) by two weeks. The latter, which seems most appropriate, runs the risk of reducing slack that may be needed later. In general, you can do the following to remove resource constraints when only one project is involved:

1. Identify resource requirements for tasks on the critical path.
2. Add resource requirements for other tasks, using desired start dates.
3. Compare resource requirements with resource availability.
4. Identify options to remove resource conflicts that are found. You may verify that the conflict is real, adjust start dates for tasks with slack, give the schedule more attention to reduce downtime inefficiency, improve productivity (with new tools, improved match of people with tasks, or incentives), adjust resource availability temporarily (by planning overtime, rescheduling vacations, obtaining temporary employees, or subcontracting tasks), change resource availability (by training or promoting surplus skill to fill needs or hiring new people), change the time schedule (accelerate the

A project schedule that requires use of already assigned resources is unrealistic.

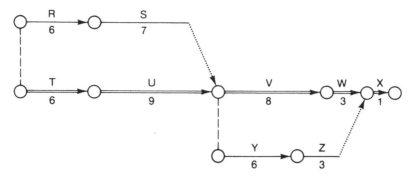

FIGURE 10-4. *Resource allocation with second project. (This project starts at the same time as the first project. The network diagram, including task durations, is shown.)*

critical path, which will put you ahead of schedule, delay the critical path, which will make you late, or change the start date of new projects), or change the plan (specifications, task sequence, or standards).

Suppose that you have a simultaneous second project (Figure 10-4). Suppose that task W requires four inspectors and you have only six. That's no problem on the second project taken alone, but it is when you try to accommodate its need for inspectors with the first project's need, as shown in Table 10-1. Both task W in the second project and Task K in the first project are on the critical paths, and they need a total of seven inspectors at the same time. Something has to be done.

The procedure in this case is the same as removing conflicts on a single project, except that:

1. Projects must be ranked by priority.
2. The highest priority project gets the first claim on available resources.
3. The second priority project gets second claim, and so on.

> **Lower priority projects usually lose the competition for limited resources.**

The result is that lower priority projects usually get delayed, performance is compromised, or there is a lot of (unplanned) subcontracting. Sometimes it is cost-effective to accelerate a small or low priority project to get it out of the way and thus avoid a major conflict that would otherwise arise.

Project schedules are usually prepared initially without regard to whether the required resources will actually be available when desired or required. Thus, there can be a serious problem if the impact of resource constraints is overlooked. The first step to avoid this problem is to refine the schedule for your project so that all tasks are consistent with available resources. Then the resource requirements of other projects must be checked and conflicts resolved. These other projects include both those that might start and be in process during your project and those existing (or planned) projects that are supposed to be completed before your project starts but that are delayed to impact your project. If this is not done, the lack of resources will not magically cure itself; it will become an obstacle when there is less (or no) time to devise an alternative schedule.

COMPUTER SOFTWARE

Most computer software, including Microsoft Project, will produce resource histograms. Figure 10-5 is the TSTETIL diagram for a simple project (project A) that requires two resources (alpha and beta, which might be people having a specific skill or a facility dedicated to a specific purpose, such as a high temperature weatherometer). Figures 10-6 and 10-7 are the resource histograms for these two resources, neither of which is overcommitted.

A network diagram is not merely a schedule dimension plan; it also clarifies resource allocation.

Figure 10-8 is the TSTETIL for another simple project (project B), and visual inspection of the figure reveals that resources are not overcommitted on it when it is considered in isolation. However, when projects A and B are both run as scheduled using the same resource pool, as shown in Figure 10-9, resources alpha and beta are overcommitted, as shown in Figures 10-10 and 10-11. Resource gamma, shown in Figure 10-12, is not overcommitted.

Many computer software packages, including Microsoft Project, will also automatically level resources or resolve resource overloads. However, the software algorithms do not necessarily produce a practical schedule. Therefore, this resource level capability should be tested and carefully evaluated before you apply it on your project.

TIME VERSUS COST TRADE-OFF

CPM has historically been associated with network diagrams in which there is considered to be a controllable time for each activity. This implies that activities can be accelerated by devoting more resources to them. Thus, there is a time versus cost trade-off for each activity and consequently for a path or the entire project.

Figure 10-13 shows this kind of situation. If you are trying to accelerate a project, you should accelerate the critical path. Of all the activities on the critical path, the most economical to accelerate are those with the lowest cost per amount of time gained.

Consider the following situation that you might face as the project manager for the project illustrated in Figure 8-10.

To carry out this project (assuming all times are in weeks), you must rent a standby electrical generator for the entire duration of the project (however long it is) at a cost of $1,000 per week. Your purchasing department has told you that the subcontractor

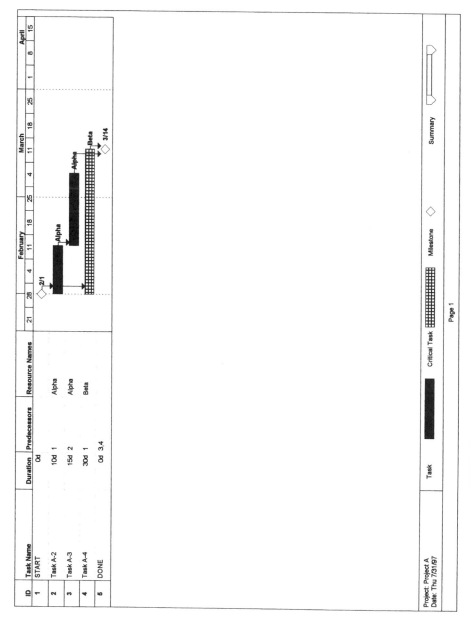

FIGURE 10.5. *TSTETIL schedule for project A showing required resources.*

138

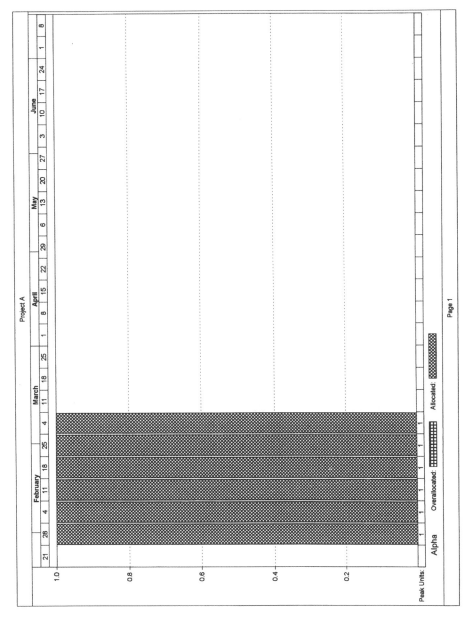

FIGURE 10-6. Resource histogram for resource alpha for schedule shown in Figure 10-5.

FIGURE 10-7. Resource histogram for resource beta for schedule shown in Figure 10-5.

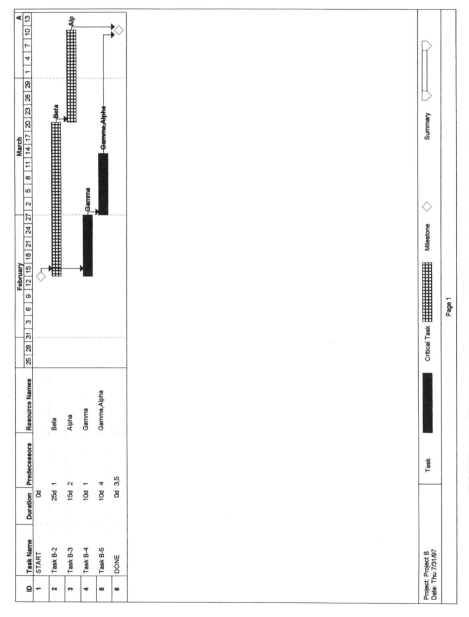

FIGURE 10-8. TSTETIL schedule for project B showing required resources.

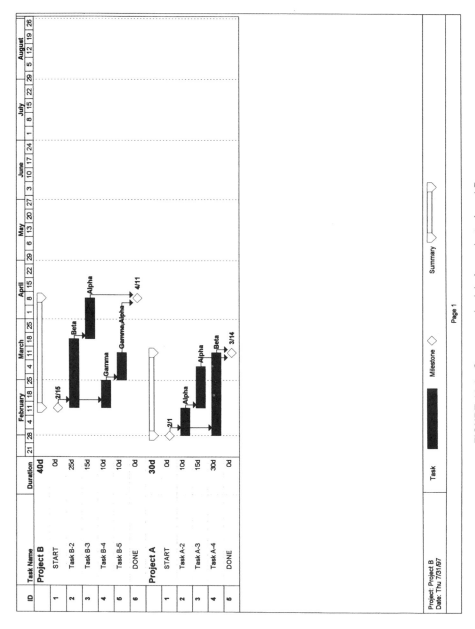

FIGURE 10-9. *Common schedule for projects A and B.*

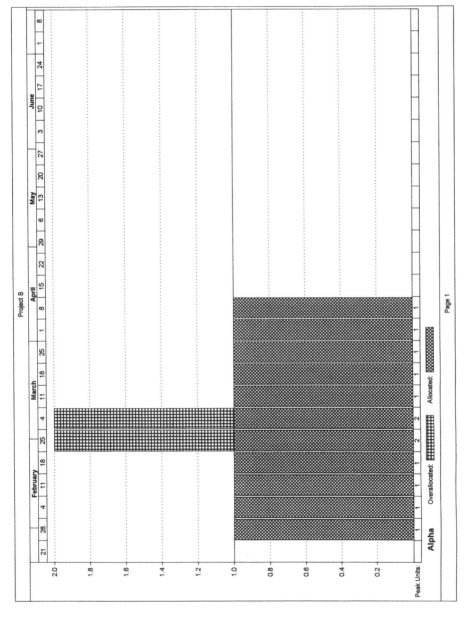

FIGURE 10-10. Resource histogram for overcommitted resource alpha for schedule shown in Figure 10-9.

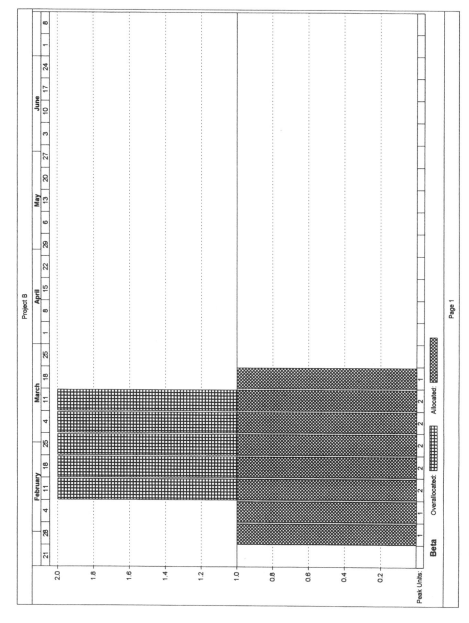

FIGURE 10-11. *Resource histogram for overcommitted resource beta for schedule shown in Figure 10-9.*

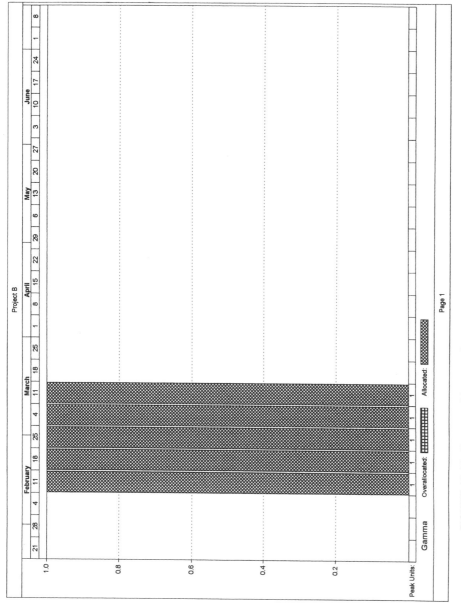

FIGURE 10-12. Resource histogram for resource gamma for schedule shown in Figure 10-9.

145

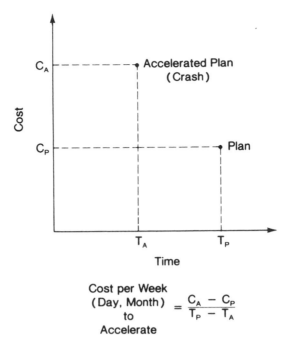

$$\text{Cost per Week} \atop {\text{(Day, Month)} \atop {\text{to} \atop \text{Accelerate}}} = \frac{C_A - C_P}{T_P - T_A}$$

FIGURE 10-13. *Time versus cost trade-off.*

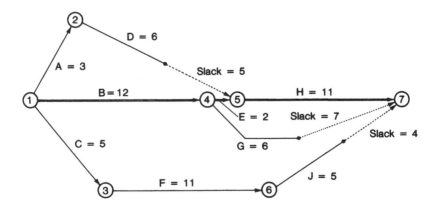

The saving of $200/week can be obtained
for only four weeks, after which path
C-F-J also becomes a critical path.

FIGURE 10-14. *Time-based activity-on-arrow diagram showing that the saving of $200/week can be obtained for only four weeks, after which path C-F-J also becomes a critical path. (Depending on the certainty you feel for C-F-J and B-E-H, it might be better to shorten only activity B by three weeks, thus maintaining only one critical path.)*

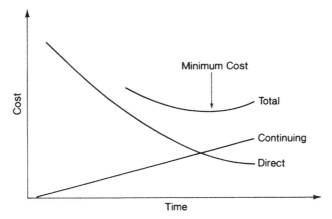

FIGURE 10-15. *Finding the lowest cost.*

performing task B has offered to shorten its performance time by as much as five weeks (that is, to seven weeks) but will charge $800 per week for every week less than the original twelve weeks (that is, a premium charge of $800 for eleven weeks delivery, $1,600 for ten weeks delivery, or $4,000 for seven weeks delivery). You can save $200 per week by accepting the subcontractor's offer. However, you have to look at your network diagram before rushing to accept the offer (Figure 10-14).

As you can see, once task B is shortened from twelve to eight weeks (at a cost premium of $3,200, producing a $4,000 saving on the standby generator), there is a second critical path. Thus, you cannot advantageously shorten task B by five weeks, only by four. In fact, you might prefer to shorten task B by only three weeks to avoid having two critical paths.

Figure 10-15 shows another aspect of this. The direct cost curve depicts those costs associated with carrying out the project that are time dependent and for which there is a cost premium associated with shortening the program. In addition, there might very well be continuing costs associated with the program, for instance, the rental of standby power generators or other such facilities. In this kind of situation, there will be a time that leads to the lowest cost for the project.

TYPICAL PROBLEMS

A major resource problem occurs when project schedules change. This may be caused by a change in your project or another project or by a shift in the start date of a project. As I discussed in Chapter 3, you can never be certain which proposals will be approved or when they will start. Therefore, you must be constantly alert for these changes and their impact on the resources you plan to use.

(continued)

When first starting a project, the project manager will devise a project plan that details the tasks and milestones along the path to completion. Time estimates can then be placed on each activity to determine the project completion date. These estimates are based on the assumption that the resources—primarily people in the organization—will be available both at the beginning of and throughout the specified tasks. However, as time and the project progress, priorities shift within the organization. When that happens, promised resources are usually moved on to higher priority projects (which often means those that are in bigger trouble). This then results in a change in the project's completion date. Project managers have to anticipate and plan for such resource deficiencies. Otherwise they will often be in the uncomfortable position of trying to justify a schedule slip to the sponsor when it was not due to any sponsor action.

Another problem can occur when you wish to use help from some other function within your organization. Imagine that your project needs help from functions A, B, and C (whatever these departments or divisions may be). The current workload forecasts for those functions are shown in Figure 10-16. The

(continued)

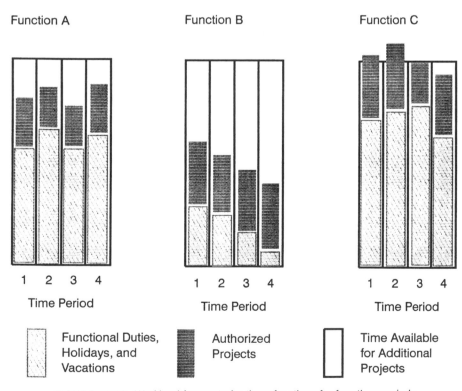

FIGURE 10-16. *Workload forecasts for three functions for four time periods.*

time periods might be days, weeks, months, quarters, or years. If the work you have for each other function demands exactly the amount of uncommitted time shown for function A, you should anticipate the following reactions. Function A will be able to help, and your requirements will exactly (and helpfully) fill up their unutilized time. Function B is likely to *really* "help" you, saying that you don't truly understand the task's difficulty. They will try to persuade you that the task requires far more work than you thought, and they are likely to try to inflate your needs to fill up all their unutilized time. Function C will be unwilling even to talk with you because they already have too much work. Therefore, understanding the workload in other functions can be important if you need their assistance.

There are also several subtle problems for which you should be alert. Finding highly qualified team members is always a problem because there are never enough. You will often have to use what's available rather than what's ideal. Breaking any resources loose from their last project always takes more time than anticipated. Getting appropriate time from individuals who have only a part-time assignment to your project can be a challenge. When confronted with this, try to schedule full days rather than parts of days.

Computers are very helpful for manipulating and summing personnel resource requirements on many simultaneous projects and thus can help alert you to prospective resource constraint problems. When computers are used to assist with planning (or anything, for that matter), there is always the danger of entering incorrect data or making programming errors. With standard project management software, the software is proven, but there can still be incorrect data entry. To avoid this kind of problem, verify data entry or run manual spot checks of output. Never blindly accept any computer output as gospel.

Month & Hours Allowed / Task or Project	1	2	3	4	5	6
	160	160	200	160	160	200
Materials Study	100	4	2	60	80	100
Ink Development	20	60	80	60	40	
Quill Development			40	40	40	
Chemical Society				40		
Total Hours	120	64	122	200	160	100
Available or ⟨Conflict⟩	40	96	78	⟨40⟩	—	100

FIGURE 10-17. Projected six-month workload for individual, showing conflict in month 4.

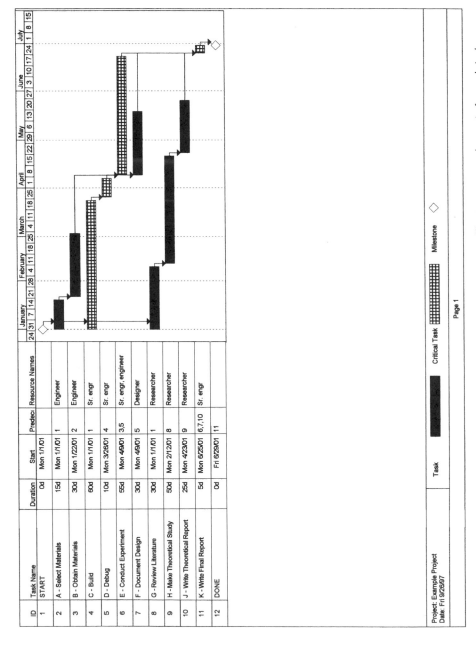

FIGURE 10-18. TSTETIL schedule for the example project showing specific resource requirements for each task.

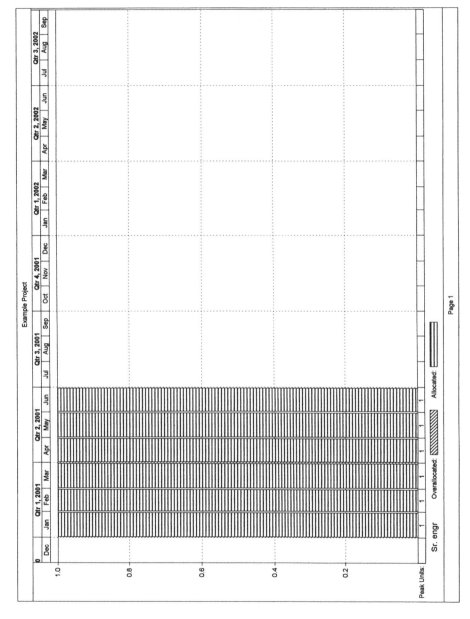

FIGURE 10-19. Resource histogram for senior engineer for example project schedule shown in Figure 10-18.

151

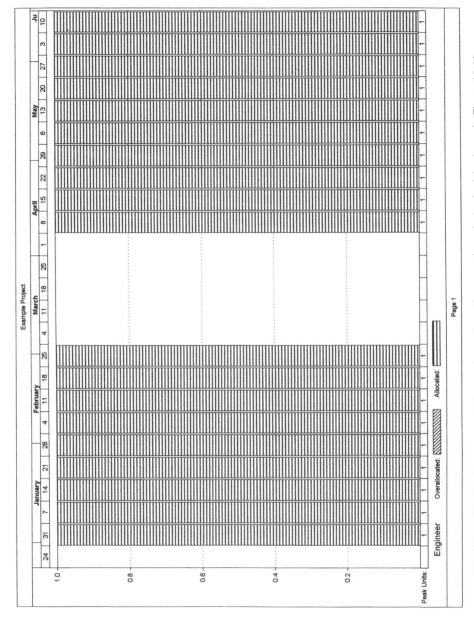

FIGURE 10-20. *Partial resource histogram for engineer for example project schedule shown in Figure 10-18.*

FIGURE 10-20. (*Continued*).

MATERIALS STUDY PROJECT

The materials study project can be used to illustrate a resource allocation problem. Cheryl Williams is slated to work on three projects and also plans to attend a society meeting. A workload forecast for her (Figure 10-17) would show the conflict. Once it is identified, corrective action could be initiated (such as planning overtime work, canceling her society participation, or finding some other chemist to work on the quill development project).

Figure 10-18 is the TSTETIL for the example project, showing the specific resource categories required for each task. This is similar to Figure 7-19. Figures 10-19 and 10-20 are the resource histograms for two of those resource categories.

In more complex situations, where there are many projects, holidays, and vacations to be considered, a computer is usually better able to handle the data. Project management software often includes priority scheduling in which low priority projects are rescheduled so as not to conflict with high priority projects (which usually results in low priority projects being late). "What if" capability is also often included, allowing managers to explore all the consequences of various scheduling options.

HIGHLIGHTS

- Resources, whether people or things, should be carefully allocated in a project.
- A network diagram can clarify resource allocation.
- Each activity, critical path, and project has a time versus cost trade-off.
- Computers can aid project planning in several ways, although care must be taken to avoid entering incorrect data and making programming errors.

11

Risk and Contingency

In this chapter, I review the sources and residual amounts of risk in projects. Then I discuss contingency, which is a trade-off with risk, and describe ways to insert contingency and reduce uncertainty. Some ways in which computer software can help are also explained.

RISK

Projects contain inherent risk. Table 11-1 identifies some of the factors that lead to lower or higher risk. This risk has to be recognized and reduced to the extent possible or practical, and the residual risk must be managed. Figure 11-1 illustrates that if the performer does not have any contingency (discussed later in this chapter), the performer will bear all the inherent risk. If the performer inserts infinite or a very large amount of contingency, the sponsor will bear essentially all the risk—but the sponsor is unlikely to authorize such a project. Thus, realistic contingency allowances have the effect of allocating the residual risk between the sponsor and performer.

One approach is to make multiple time (and cost) estimates for each task. Then the project manager and task managers can decide which estimate to present in selling the project. This is always a trade-off with risk (see Figure 11-2). The shorter the proposed time schedule (or the lower the bid price), the greater the risk of time (or cost) overrun.

Risk actually has two components: tangible (for instance, financial) and psychological. The latter is often more important because some people have a very low tolerance for tangible risk and thus insist on padding their schedule and budget estimates. Many technologists (such as engineers and programmers), educated in the precision of numbers, hate to make a numerical estimate (time or money) that may later be proven wrong; therefore, they have a low psychological tolerance for making optimistic time and

Risk has both tangible and psychological components.

TABLE 11-1   Some Elements of Project Risk

Factor	Lower Risk	Higher Risk
What is to be produced	Similar to prior work	Highly novel
How it is to be made	Well known	Unknown
Duration of project	Short (less chance for the world to change)	Long (global conditions may change) or very short (no time to recover from problem)
Prior experience with sponsor	Successful	Hostile or adversarial
Importance of project to your organization's management	Top priority	Unimportant
Availability of skilled resources	Plentiful	Some required skills missing or otherwise committed
Background of project and support teams	Pertinent experience and have previously worked together	Lack relevant experience and cooperative work history
Attitudes and beliefs of project and support teams	Convinced Triple Constraint can be achieved	Have reservations about achieving Triple Constraint or are convinced Triple Constraint cannot be achieved
Sources of critical components or materials	Multiple, reliable sources	Only one source of uncertain reliability
Project's reputation (if it has been running for some time)	Great	Very bad

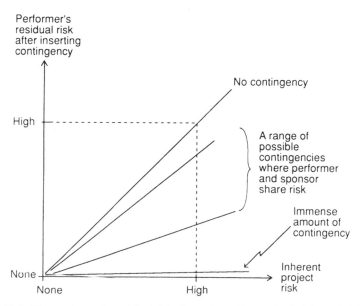

FIGURE 11-1. *The performer's residual risk depends on the project's inherent risk and the amount of contingency in the performer's plan.*

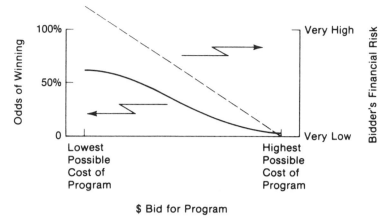

FIGURE 11-2. *Contingency and risk trade-off.*

cost estimates. Such a psychological bias induces technologists to make pessimistic time and cost estimates so that they can then do better. (However, this bias does not prevent naivete from producing optimistic estimates.)

A more senior manager has a different perspective. He or she is likely to have several project or task manager subordinates. Thus, the senior manager can average his or her risk over several projects or tasks. That is, one project or task may overrun, but there is likely to be one with a compensating underrun.

In general, when two people are in a dispute or argument about the schedule or budget for a task or project, the point of contention is not the estimate per se; it is about the risk. Unfortunately, the energy of the argument is usually concentrated on whether the work should take, for example, four or six weeks (or cost $4,000 or $6,000), which cannot be known with certainty in advance. It would be far more productive to discuss what the risks are if the proposed schedule is planned for four weeks rather than for six weeks.

One approach to risk analysis is illustrated in Figure 11-3. This kind of chart could be annotated by the task manager when he or she is meeting with the project manager and others to make the task's time and cost estimates. A large contingency (or, better, a conservative backup alternative) is required if there are any possible problems in the upper right area (1 in Figure 11-3). The lower right (3 in Figure 11-3) deserves a small inflation of each affected task's schedule and budget. The lower left (4 in Figure 11-3) can be ignored. The zone of greatest difficulty is the upper left (2 in Figure 11-3) because it is hard to envision a low probability event with a severe impact. This can best be handled as part of the general contingency that is inserted.

CONTINGENCY

Plans represent the future. Because nobody has a crystal ball, plans must include contingency. Contingency can be thought of as the antidote to risk. As illustrated

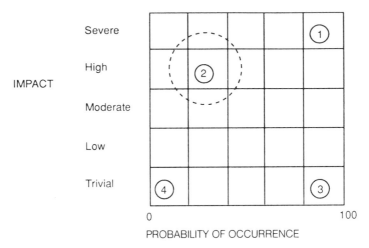

FIGURE 11-3. Risk analysis chart.

This material originally appeared in a videotape presentation produced by the Association for Media-based Continuing Education for Engineers, Atlanta, Georgia.

in Figure 11-1, the performer who can insert a lot of contingency will bear very little residual risk. In fact, this contingency should be placed on each of the three dimensions of the Triple Constraint. Smart customers know that contingency is required and often want to know how much and what kind is included—not to try eliminate it, but rather to be assured that their own plans (which include your project) can be achieved. (Unfortunately, some government procurement regulations prohibit or inhibit this realistic approach, which may contribute to the costly overruns frequently reported.)

> **Build contingency into all project plans.**

On the performance dimension, it is important that the contingency not take the form of gold plating. Where appropriate, include a small design margin. For instance, if the goal is to have a new car's weight be only twenty-five hundred pounds, it might be appropriate to try to design the car to weigh twenty-three hundred pounds. However, never carry this to an extreme (targeting the car's weight at fifteen hundred pounds, for example).

Contingency is most often associated with the schedule and cost dimensions because projects will inevitably encounter difficulties there. Many things that occur simply require more time and money than planners think. For instance, whenever you must interact with other people, obtaining your boss's approval, perhaps, their schedules constrain you. You will not have instant access, and a delay will occur. Customer furnished items, such as interface hardware or sample data sets for computer programming projects, frequently do not arrive when expected or in the condition promised. Work done at remote locations often takes longer than work done in your own organization's facilities because things like copying machines or other support resources are simply not as conveniently available. If your project involves hiring people, it takes time to train them and make them effective workers on the job. Even if you don't have to hire people, you must still indoctrinate the people who are assigned to work on your project. Similarly, there will be illnesses and vacations pulling people away from the job. These and other tasks, some of which are included in the following list, make it important to build in schedule and cost contingency.

Resource availability

Resources conflicts

Motivation

Something else becomes higher priority

Interface with others

Miscommunications

Get approvals

Get support from other groups

Place major subcontracts and purchase orders

Subcontractor delays

Adjust for labor strikes

Make mistakes

Inefficiency

Train people

Replace sick and vacationing personnel

Cope with personnel resignations

Obtain security clearances

Obtain customer furnished equipment and information

Work at remote locations

Currency rate fluctuations

Overhead and other rate changes

Cope with travel delays

Handle customs duty clearance

Comply with customer procedures

Determine whether a new design will work adequately

Advance the state of the art

Accommodate computer downtime

Ways To Insert Contingency

There are several ways to insert time and cost contingency. The first is to have everybody who provides an estimate make his or her own time and cost contingency estimates. The problem with this approach is that contin-

Beware of applying contingency upon contingency.

gency gets applied on top of contingency, which is then applied on top of other contingencies, and so on. It does not take many multiplications of 110 or 120 percent before the price of the entire project exceeds the sponsor's reach, and it does not take many extra hours, days, weeks, or months before the schedule becomes unreasonable.

The second method to insert schedule and cost contingency is to put a small amount of contingency, 5 or 10 percent, on each activity in the network. This method is fine, but it misses the point that some activities can be accurately estimated and some others cannot.

A third method to add contingency is to add an unplanned (and, it is hoped, unrequired) activity. For instance, adding a system management task for the duration of the project is really a variation of the second method but provides only cost contingency. A given level of cost contingency, the amount of the system management task, is attached to the entire project. A weakness of this method is that it does not inherently provide a time cushion as well.

A fourth method is a variation on the third method. This fourth method, which I prefer, is to add one or more additional plausible (but, it is again hoped, unrequired) tasks near the end of the project, as shown in the middle of Figure 11-4. This has the effect of pushing activities forward from the project's scheduled completion, in this case, shipment, to the earliest possible point. These extra tasks thus have the effect of providing schedule contingency. Because money is devoted to these extra tasks (not necessarily in direct proportion to their increased time), they also supply a cost contingency. However, both the time and money contingencies are purely arbitrary.

A fifth method is a variation of the second method (that is, adding 5 or 10 percent). The variation, which is far better than the second one, is to explain to everyone providing estimates that they should be as accurate (or optimistic or pessimistic) as possible. Then the entire group can discuss how much schedule and cost contingency should be put on which activities. This can be done by considering the likelihood of things going wrong, the importance of such an outcome, and the maximum and minimum impacts produced by this undesired outcome. There might be some highly uncertain activities that receive a contingency of 50 percent or more. Conversely, a final report might be assigned no or only a small percentage contingency. This fifth method is illustrated in the bottom of Figure 11-4.

Ways To Reduce Uncertainty

Probably the best way to reduce uncertainty is to break the entire project into many small phases, the first of which is a study or definition task. The goal of that first phase is to reduce uncertainty for the remainder of the project. You may also find it helpful to make models, prototypes, breadboards, or other demonstrations. Computer-generated three-dimensional (3D) images can be viewed and discussed by groups of team members. These 3D images can be rotated and otherwise manipulated, and changes can be tested on a "what-if" basis. With suitable equipment, solid models can be constructed

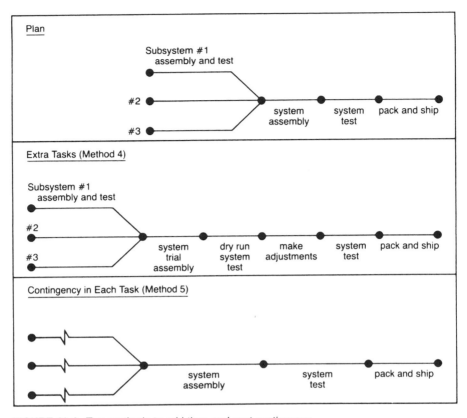

FIGURE 11-4. *Two methods to add time and cost contingency.*

This material originally appeared in a videotape presentation produced by the Association for Media-based Continuing Education for Engineers, Atlanta, Georgia.

in a variety of plastic materials (so-called "rapid prototyping"), providing a way to "touch and feel" the item. Modern prototyping tools for software development or computer-aided design and engineering software packages may also be helpful.

Testing is very important in reducing risk. Nonworking models may be tested for form and fit and examined for style and other industrial design aspects. Working prototypes of components, products, or processes can be tested for whichever functions are embodied. If the test results are unsatisfactory, either the performance goal or design must be modified. A finished product may be tested for reliability, lifetime,

Plan for and conduct appropriate tests to reduce uncertainty and risk.

and varied performance and use aspects (for example, regulatory and standards compliance, customer and user satisfaction, and so on). Figure 11-5 shows the impact on the schedule of too little and too much testing: Problems may go undetected if too few tests are performed; excessive tests extend the schedule. It is also vital to plan for and assure that any special purpose test equipment will be validated.

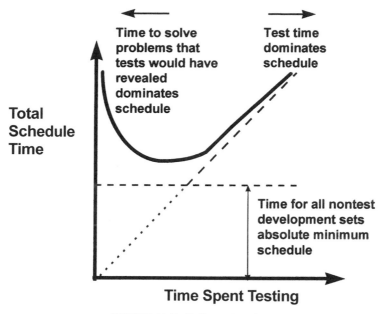

Time Spent Testing

FIGURE 11-5. *Optimum test time.*

In the case of software, especially safety-critical software, the quality assurance standard for the test software must be as high as that of the software being developed.

In addition, using experts or people with relevant experience can be tremendously helpful. Similarly, assuring yourself as project manager that you are the single focal point with total responsibility can reduce communication problems, which are often the cause of major uncertainties.

Finally, if the major uncertainties relate to schedule, you can help yourself by starting early. You may also wish to assign additional resources to activities on the critical path. In the case of tasks that have great technical difficulty, you may wish to run a second approach in parallel. Finally, you must of course be certain that everyone paid a great deal of attention to the schedule.

COMPUTER SOFTWARE

As I said before, some project management software programs can control graphic output terminals to produce network diagrams or bar chart displays. Computers can also furnish summary project planning and control information for several projects and can easily provide a wide variety of other planning detail to the project manager, support group managers, and task leaders. Thus, it is easily possible to make several different time and cost estimates for each project task, where different amounts of contingency and risk are involved. This can provide a summary overview of the project under these different assumptions. The resulting computer output then allows a really useful discus-

> *Computers are often useful for contingency planning.*

sion. Rather than have a senior manager (or customer) say the project is too long or too expensive, they can examine the greater risks attendant to settling on a shorter schedule or lower budget.

TYPICAL PROBLEMS

> The worst problem concerning contingency is that some managers and customers believe they can save time and money by eliminating it. This belief, or pretense, that contingency is not required restricts the project manager's ability to cope quickly, efficiently, and effectively with the problems that inevitably arise on a project. The lack of contingency reserve (especially time and money) means that the project manager must have a customer or management negotiation every time a problem arises.
>
> Before the fact, it is impossible to state accurately how long a task or project will take or how much it will cost. Consequently, talking about the estimated amount of time or money is not helpful. Instead, it can be very helpful to discuss the likelihood that the estimated time and cost can be met and upon what factors that likelihood is contingent. Such a discussion of likelihood can clarify assumptions and the risk in them and may point to an alternative, lower risk plan.

MATERIALS STUDY PROJECT

> Table 11-2 illustrates the various time estimates that might have been made under different risk assumptions. Note that the critical path is changed when the pessimistic estimates are used because tasks A and B are much longer.

TABLE 11-2 Multiple Time (and Cost) Estimates, Using Network Diagram of Figure 10-2

Task	80% Optimistic	50-50 Plan	20% Pessimistic
A	2	3	6*
B	5	6	14*
C	11*	12*	14
D	2*	2*	4
E	10*	11*	13*
F	5	6	8
G	6	6	8
H	10	10	12
J	5	5	6
K	1*	1*	2*
TOTAL	24	26	35

* = Task is on critical path.

HIGHLIGHTS

- All projects should contain contingency, which may be best inserted by adding tasks near the end or by distributing it in each task.
- The amount of contingency is a trade-off with risk.

Part 3

Leading the People Who Work on a Project

12

How to Organize a Project

In this chapter, I describe the three main forms, namely, functional, project, and matrix, by which organizations arrange their internal reporting relationships and chains of command. Any of these three forms may also be used for a project within an organization. Then I describe four other forms by which projects may be managed. Finally, I discuss the informal organization.

THREE PRINCIPAL ORGANIZATIONAL FORMS

Projects have a finite life, from initiation to completion. Conversely, a company, government department, or other organization expects to exist indefinitely. This temporal difference makes it difficult to organize and manage a project within a larger organizational entity.

In addition, projects frequently require the part-time use of resources, but permanent organizations try to use resources full-time. Typical project requirements include the following: one hour of computer time each day for a week; use of a backhoe next Tuesday for the afternoon; one-quarter of Jane Draftperson's time this month and three-quarters of her time next month; use of Joe Technician full-time as soon as the project's circuit designer completes the design. No economically viable organization can afford to stockpile these resources to serve any project's needs instantly. Thus, it is important to organize for project work in adequately responsive ways, and it is important for project managers to recognize that this is a compromise that is not fully responsive to project needs.

Although no organizational form is perfect, it is important to support the existence of projects when they are present. This means the organization must plan to accommodate this temporary disturbance and accept some disharmony.

There is a variety of ways that companies or their divisions or governmental organizations can be organized and effectively manage projects. The three most common of these organizational forms are functional, project, and matrix.

Organizational forms differ in response to projects.

FIGURE 12-1. *Typical organization chart of a functional organization.*

Functional

Functional organizations (Figure 12-1) are common in companies dominated by marketing or manufacturing departments (whenever there is a large amount of repetitive work) and exist in other kinds of companies as well. The person asked to manage a project in a company with a functional organization has generally been oriented and loyal to the functional group to which he or she belongs. Specialists are grouped by function, encouraging the sharing of experience and knowledge within the discipline. This favors a continuity and professional expertise in each functional area.

> **From a project management point of view, the functional organization is least desirable.**

Because such an organization is dedicated to perpetuating the existing functional groups, however, it can be difficult for a project to cross functional lines and obtain required resources. It is not uncommon for hostility to exist between different functions; that is, there are barriers to horizontal information flow, and open channels tend to be vertical, within each function. Absence of a project focal point may trouble a customer interested in understanding the project's status, and functional emphasis and loyalties may impede completion. As a practical matter, the only real focal point is at the top, with the president or general manager, who has far more to do than adjudicate cross-functional conflicts.

The functional organization emphasizes functional skills by concentrating these in a small group. Thus, functional experts spend most of their time in proximity to and rubbing shoulders with people of similar skills. Unfortunately, this can isolate them from others with whom they must work if the project is to be successful. Many technical specialists, scientists, and engineers, for instance, often lack people skills. This may also be true of other specialists, such as computer programmers. Because each individual function is best able to communicate with itself, interfunctional communication is often impeded and cooperation hindered.

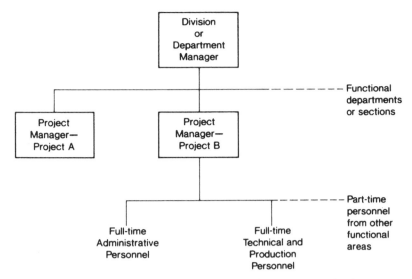

FIGURE 12-2. *Typical organization chart of a project organization.*

However, if all the project's required personnel resources (including the project manager) are located within a single functional group, many of the problems can be avoided. Thus, when all project resources are located in one group, the functional organization may be a good choice.

Project

A project organization (Figure 12-2) emerges from a functional organization when the latter impedes project needs. The solution is to move many of the people working on the project from their functional group to the project manager. Line authority for the project is clearly designated, providing a single focal point for project management. All full-time personnel are formally assigned to the project, thus assuring continuity and expertise.

A major difficulty with this kind of organization is the uncertainty these people feel about where they will go when the project is completed. This terminal anxiety can impede the project's completion. There is also a tendency to retain assigned personnel too long. In addition, it is a rare project that actually has all the required resources assigned to it. Thus, such an organization still requires the project manager to negotiate with the remaining functional organization for much of the required support.

If the organization develops additional projects, managing them in this way leads to a splintering, with many separate project centers existing apart from the functional organization. Duplication of facilities and personnel can result. Managers within the functional organization may feel threatened as people are removed from their functional group. This produces another series of stressors. Project organization

The project organization form is most useful on large projects of long duration.

often inhibits the development of professional expertise in functional specialties and may not effectively utilize part-time assistance from them.

The project manager must be a senior person (a so-called "heavyweight" project manager) to be effective in the project organization. Because he or she has direct responsibility for the work of everyone involved in the project, the project manager requires direct supervision of these people. Consequently, project managers in this organization are often of equal or greater rank than the functional managers. This can exacerbate organizational disharmony.

Matrix

The matrix organization (Figure 12-3) is a hybrid that may emerge in response to the pressures resulting from inadequacies with a functional or a project organization. It attempts to achieve the best of both worlds, recognizing the virtues of having functional groups but also recognizing the need to have a specific focal point and management function for each project. Line authority for the project is clearly designated, providing a single focal point. Specialists, including project managers, are grouped by function, encouraging the sharing of experience and knowledge within the discipline. This favors a continuity and professional expertise in each functional area. The matrix organization recognizes that both full-time and part-

FIGURE 12-3. Typical organization chart of a matrix organization.

time assignment of personnel are required and simplifies allocation and shifting of project priorities in response to management needs. Functional departments are responsible for staffing, developing personnel, and assuring the technical quality of the work done by those personnel. The project managers are responsible for defining the work to be done and establishing a reasonable plan (including schedule and budget) for accomplishing it. Project managers and department managers must jointly agree on task, goals, and the specific schedule for each of these.

The main drawback is that a matrix organization requires an extra management function (namely, project management); so it is usually too expensive for a small organization. It is even possible to have a matrix organization within a matrix organization (for example, the matrixed engineering department). In addition, the extra functional unit (that is, project management) can proliferate bureaucratic tendencies, and the balance of power between project management and functional units can exacerbate conflicts.

> *The matrix is probably the best organizational option if you have many projects.*

A matrix organization may be either weak or strong, depending on the project manager's power compared to that of the functional managers. The weak matrix may operate somewhat like a functional organization, and the strong matrix may operate somewhat like a project organization. This project manager's power may derive from financial control, seniority, or simply his or her persuasiveness. Thus, it is possible to have both a weak and strong matrix within a single organization simultaneously.

Project managers in a matrix organization are often concerned about the work of tasks being done by people in functional organizations. Obviously, project managers wish to allow and encourage these people to have as much autonomy as possible to promote synergism, which should enhance the entire organization's efficiency and quality of work. This is particularly true where the functional groups have worked well in the past and have specific knowledge and expertise relevant to the task at hand. Conversely, the project manager is responsible for the ultimate result, must integrate each functional group's contribution, and thus wants to retain as much control and influence over each functional group as possible. Striking a satisfactory balance between these conflicting objectives is very difficult.

Operating in a matrix organization can be challenging when responsibility for a project or its key parts is divided or roles are not clearly defined. Project managers in this situation must get prompt resolution from management before the work direction is confused or the schedule is delayed. Such resolution may itself be contentious because of the inherent tension between the management of project and functional departments. As an example, one project manager I met found himself in conflict with an account manager, who was generally charged with responsibility for maintaining harmonious relations with and obtaining more business from the customer's company. The account manager wanted extra unpaid work on the project for customer goodwill, but the project manager was accountable for timely completion within budget. It is clearly important to face and resolve this kind of problem early rather than hope it will happily solve itself.

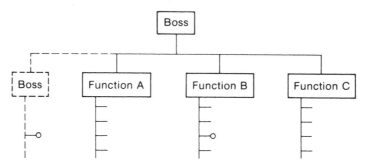

FIGURE 12-4. Quasi matrix organization.

Source: "Basic Management Skills for Engineers and Scientists." Lecture notes by Milton D. Rosenau, Jr. Copyright © 1982 by the University of Southern California. Reprinted by permission.

OTHER ORGANIZATIONAL FORMS

Quasi Matrix

The quasi matrix (Figure 12-4) is a compromise way to obtain the benefits of the matrix form in a functional organization otherwise too small to afford it. In this compromise organization, when someone is designated project manager, he or she remains part of the functional group for project work done in that group (which is the same as in the functional organization). However, for project management work, he or she reports directly to top management.

Thus, the project focus has top management's support to cut across functional group boundaries. In some cases, the quasi matrix will have a manager of projects (just as in the matrix organization), rather than the boss, to whom the project managers report for project (as distinct from functional) matters.

Unlike the "heavyweight" project manager required in the project organization, a "lightweight" project manager may suffice in the quasi matrix. In many situations, the role can be filled with a middle or junior level person, often a technical specialist (such as a design engineer, programmer, and so on). Many of the project contributors continue to report to their functional managers, and the project management role may primarily entail coordination.

Venture

The venture organization, common in several very large, commercially oriented companies, is especially appropriate for projects aimed at new product development. Basically, the goal is to set up a tiny functional organization within a giant corporation, thus achieving the advantages of compact size, flexibility, and the entrepreneurial spirit of the small company within and supported by the financial, physical, and human resources of the larger company. Where such a management organization exists, it is common to team up an engineer with a marketing person and a manufacturing person in the earliest phases of new product development. As the effort moves forward, the venture organization grows, ultimately becoming a functionally organized division within the parent company.

Task Force

Organizations frequently use a task force to cope with an unexpected project. Hence, this response is most commonly utilized by a functional organization because the other organizational forms are already able to deal with projects. A task force may be thought of as a rarely used single project organization within the functional organization. It can be formed quickly, usually by a very senior officer.

A task force may be helpful in coping with an unexpected project.

Although the people selected to serve on the task force may be highly motivated by their selection, they frequently are not relieved of their usual duties and thus may not have sufficient time for the task force. If they are relieved of their normal duties, they may be anxious about their assignment when the task force has completed its job.

Other Project Teams

Today other labels are applied to various special purpose project teams. These include core teams, concurrent engineering teams, reengineering teams, benchmarking teams, process development teams, and SWAT teams. I am sure other labels will be in vogue in the future. In most cases, such a label merely describes the role of an ad hoc project team.

Some teams may be self-managed, although this is often more in the label than in reality. Assignment to such an alternative project team may be full- or part-time; the latter is more common for all but emergency situations.

THE INFORMAL ORGANIZATION

Regardless of the formal organizational structure, there is always an informal organization (Figure 12-5). Where friendships and common interests are present, these create channels through which information flows easily and cooperation is encouraged. Conversely, rivalries and animosities can inhibit cooperation.

TYPICAL PROBLEMS

Each organizational form has its advantages and disadvantages. The only real problem occurs when a project manager believes that a different organizational form will solve all the organizational problems he or she is experiencing. In actual fact, no organizational form is perfect for all situations or for all time. Thus, reorganization is regularly required in any organization doing project work to keep the organization as useful and productive as possible.

(continued)

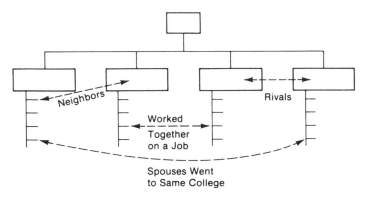

FIGURE 12-5. The informal organization.

Source: "Basic Management Skills for Engineers and Scientists." Lecture notes by Milton D. Rosenau, Jr.
Copyright © 1982 by the University of Southern California. Reprinted by permission.

FIGURE 12-6. SUPROMAC organization chart.

The persistent problem is carrying out any project that crosses the lines of more than a single functional department. In these cases, the project is likely to be at the mercy of the functions. As one project manager stated, "Each functional department determines its own priorities and staffing needs." When a project is suffering because of this parochial attitude, the manager must first attempt to resolve the problem directly with the department manager. Failing that, he or she must bring the problem—and some proposed solutions—to the attention of senior management.

MATERIALS STUDY PROJECT

The organization chart for SUPROMAC is illustrated in Figure 12-6. This is a functional organization. Therefore, Mel Chase will probably have problems getting support from other groups and sections unless a quasi matrix is established.

HIGHLIGHTS

- Three common organizational forms for project management are functional, project, and matrix.
- Four other forms that may be used are quasi matrix, venture, task force, and ad hoc project team.
- The informal organization is always present and affects how well the formal organization works.

13

Organizing the Project Team

The project team consists of those who work on the project and report administratively to the project manager. This is in distinction to the support team (people who work on the project but do not report administratively to the project manager), which I discuss in Chapter 14. In this chapter, I first review sources of project personnel and consider the frequent necessity to compromise by using whoever is available. Then I deal with how much control a project manager can exercise over project personnel and provide some practical tools to help him or her gain effective control. The next section discusses the use of task assignments both as a means to assign the work packages and to obtain commitments from personnel to carry out the work. The last section deals with using computer software in team organization.

DEGREE OF ASSOCIATION WITH THE PROJECT

Table 13-1 shows eight categories of personnel assignment to projects. They result from all possible combinations of three factors: (1) whether personnel report directly to the project manager or are administratively assigned to someone else, (2) whether they work full-time or only part-time each day (or week or year) on the project, and (3) whether they work on the project from its inception to completion or for only some portion of the project.

Project Team

The project team is composed of the people who report administratively to the project manager (the four cells so designated on the left side of Table 13-1). I consider this the project team because the project manager can assign work packages

TABLE 13-1 The Project Team and the Support Team

		Reports to Project Manager		Works on Projects but Reports to Another Manager	
Duration of Project Assignment	Nature of Reporting Relationship	Works Only on Project	Also Has Other Assignment(s)	Works Only on Project	Also Has Other Assignment(s)
From start to finish		P	P	S	S
Only a portion of project's duration		P	P	S	S

P = Project team
S = Support team

to these people rather than having to negotiate with other managers to obtain commitments for their work.

Team Matrix

The amount of project labor obtained from each category depends on the project contractor's organizational form (that is, functional, project, or matrix) and project size. In a matrix organization, no one may be assigned to work for the project manager; the entire labor pool may be drawn from the support team. In a pure project organization, the vast majority of project labor, perhaps all of it, may be assigned to the project manager. This is especially likely for a large project of long duration. As a practical matter, small projects are not likely to have their own personnel, regardless of organizational form. The majority of those assigned to a project from start to finish is either managerial or administrative because most other skills are required for only some portion of the project.

A key point that emerges from consideration of Table 13-1 is that the project manager must provide eight different kinds of management attention to people working on the project. People who work on the project for only a portion of its duration must be managed to be ready when needed; then orientation to the project must be provided; finally, the project manager must recognize that they may be frustrated or lack a sense of accomplishment at leaving the project prior to completion. People who have other assignments must be persuaded that the work on the project deserves

There are eight ways to assign personnel to a project; so the project manager must manage differently, as required by each way.

their attention each day (or week). If everyone who is working on the project is doing so on a part-time basis, other work (sometimes more interesting, but lower priority) intrudes. Because they may have a lower stake in the project, they often require better or more forceful leadership. People on the support team must be managed through other managers, which can lead to distorted work direction. Issues of priorities, performance standards, and loyalty often require the project manager's attention.

SOURCES OF PERSONNEL

There are many sources of people, including the proposal team, other people already employed by the organization, and people from outside the organization (hired personnel, contract personnel, consultants, and subcontractors).

The Proposal Team

The proposal team is the best source of project personnel.

By far the best source of project personnel is those who worked on the proposal. They are familiar with the subject matter of the proposal and perhaps to some extent with the customer's specific problems. Being already "up to speed," they require minimal indoctrination. They will, for example, presumably understand the meaning of potentially ambiguous words (such as "security"). In many instances, they will have already "bought in" to the project. Although the proposal team may be highly qualified, it probably consists of many fewer people than will be required to successfully complete the project.

Other Organizational Employees

Other employees of the organization are the second best source of personnel. These people are at least familiar with company policies and procedures; they know where the library, the model shop, and such are located. Although they may not be familiar with the specific subject matter of the proposal, they are at least familiar with how the company does business and know its strengths and weaknesses. They know whom to call for help and where to go to get something done. In fact, they have probably worked on similar projects in the past.

The project manager may know their strength and weaknesses and thus be able to assign them to appropriate work packages. The project manager will not ask them to do more than their capabilities permit or give them a work package so trivial as to be demeaning or demotivating.

People from Outside the Company

There is a variety of outside sources for personnel (see Figure 13-1). Consultants, contract ("body shop") personnel, and subcontractors can be obtained quickly, but it often takes months to hire a new person. The newly hired person, who can work for and be fully under the administrative control of the project manager, may be thought of as part of the project team in the sense I am using that term. To hire a person, a project manager has to have a personnel requisition approved, typically

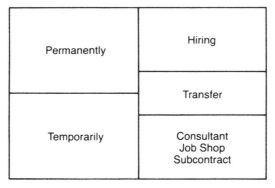

FIGURE 13-1. *Filling personnel needs.*

Source: "Basic Management Skills for Engineers and Scientists." Lecture notes by Milton D. Rosenau, Jr. Copyright © 1982 by the University of Southern California. Reprinted by permission.

must advertise the position, interview several people, make one or more offers to get an acceptance, and wait for the person to relocate (if required) before coming to work. Then there is an indoctrination period while the person becomes familiar with company practices. A project of short duration rarely can afford the time to hire personnel and thus depends on the support team.

COMPROMISE

It is indeed rare that a project manager can staff the project entirely with personnel who (1) already work for him or her, (2) worked on the proposal, and (3) represent exactly the right distribution of skills to carry out the project. Usually, the project manager must staff the project from whoever is currently available either full- or part-time. Many of these people will not completely meet the requirements. It is often a case of fitting square pegs in round holes.

Staffing compromises are usually necessary.

Qualifications

The newly appointed project manager confronted with the urgent need to staff a project team is often victimized by other managers in the company who offer their "cats and dogs." These people may be marginally employed; so company management may pressure the project manager to accept them into his or her group. Consider the problem one project manager called to my attention:

> I inherited an employee who had been with the company for approximately thirty-five years. In all but one or two years of that time he was never properly supervised. He did, however, receive regular raises. He's apparently been allowed to set his own standards of performance, but these are subpar. He has proven to be very devious at

beating "the system" with regard to hours and personal activities that are not job related. Now, my attempts at corrective actions and discussions with him produce a very defensive position. He says, "I never knew I wasn't up to par." He exhibits resentful behavior, including mistakes and sulking.

This is a very tricky situation. There is pressure from above to accept the people, and there is another manager offering them as freely available. But, if these people are known to be marginal workers, it is probably better to terminate their employment than to shift them from one project to another, burdening these projects and retaining marginal workers for long periods of time. Nevertheless, it is common for a newly appointed project manager to be offered all kinds of personnel for transfer. On a short duration project, it may be better to accept these workers, unless they are clearly unqualified, than to recruit better qualified assistance.

Motivation

Some projects that offer high pay (such as major construction projects requiring a great deal of overtime or shift premium work) frequently attract workers whose primary motivation is money. The project manager may be besieged by candidates who wish to go to work on his or her project. Their motivation, however, may not be best for the project. So confronted, the manager should seek to staff the project team with a few high quality people and confine the money seekers to support team roles, where they are someone else's problem.

Conversely, a project with high scientific content or one of national importance (such as a new medical instrument system that reliably diagnoses any cancer at a very early stage to permit effective treatment or, perhaps in the future, a manned landing on Mars) often attracts highly dedicated, altruistic people. A common correlate of this altruism is a lack of practicality, which the project manager must watch for and temper.

Some project personnel may be poorly motivated and some may be unrealistic.

Some projects have an unsavory reputation (fairly or unfairly earned) that makes it very difficult to recruit personnel. They often require portions of the work be performed at an unattractive or remote location. To overcome this drawback, various inducements may be required.

Recruiting Qualified Help

Some compromise is clearly required in staffing the project team, but there may be some skill requirements that cannot be compromised.

Most project managers prefer to have people on the project team because it seems to improve project control. People on the project team cannot be given other distracting work assignments unless the project manager approves it, but people on the support team may be given other work that detracts from their ability to honor support commitments.

Thus, in many situations where you want the best qualified help for your project, you must locate a person with the required skill and try to obtain a transfer (perhaps temporary). In some cases, the transfer request will be refused. If your organization has a matrix form, of course, transfer is not possible; at most, you can obtain a firm commitment to assign a needed, uniquely skilled person only as long as the project requires him or her.

CONTROL

Supervision

In talking about the project team, I am talking about people who work for the project manager. They may not work full-time on the project, either for its entire duration or full-time within any given workday or workweek. Nevertheless, they are under the project manager's direct supervision, unless there are intermediate levels of supervision. Some of these people may have been transferred from other managers.

Projects go through different phases, which implies that personnel must be changed. For instance, the creative design person, so valuable in the early phases of system design, is not needed when the project is moving toward completion and the team is trying to finish what has been designed rather than figure out additional clever ways to design it.

Thus, some project team people may have to be reassigned during the project. An administrator or junior project manager assigned to work under the project manager may be needed the entire time, but other personnel may need new assignments. They will either go to work on another

> **People must join and leave the project as needed.**

project full-time after completing their work or work on two or three projects part-time. It is therefore important that the project manager exercises control over the timing of these assignments so as to have people with the right skills available when required and have other assignments for them when they are not required. This is one reason a resource allocation analysis is desirable.

Proximity

One of the project manager's most powerful tools for improving control of project personnel is to locate everyone in a common area. This aids communication, and, where there is increased communication, there is increased understanding. This also improves the likelihood that everyone on the team understands the Triple Constraint. Co-location is often impractical because the people from a key required function are still needed in their "normal" (nonproject) job or some key functions are located in multiple other facilities (for example, dispersed manufacturing sites). Figure 13-2 summarizes some of the project duration and team size issues in deciding whether co-location is appropriate for your project.

> **Co-location is most appropriate for projects of long duration requiring a large team.**

Team Size	**Large**	Intrateam communication "overhead" costs are higher. An investment to promote teamwork is required. A more experienced project manager is required.	It frequently reacts to marketplace changes. Miscommunication, confusion, inefficiency, and wasted money are possible.	Team co-location is often required and justified.
	Small	Generalists may be required. A less experienced project manager may be adequate.	Cost is normally low.	It must be determined if a larger team could shorten the schedule.
			Contributions of part-time people are frequently required. Team co-location is hard to justify. Changing the project manager is normally not practical.	The project manager probably will change as phases shift. Full-time, dedicated project team members can be used. Boredom or burnout are risks. Team co-location is possible.
			Short	Long
			Project Duration	

FIGURE 13-2. *Team organizational issues affected by team size and project duration.*

Problems

Most projects cannot be fully staffed with project team members.

Most project managers would rather staff the project with many project team members and fewer support team members. However, this staffing forces the project manager to contend with many personnel problems, such as people quitting (either the company or organizational unit) to work elsewhere, sickness, higher management reassignments, lack of interest in the project, or other conflicting assignments.

TASK ASSIGNMENTS

I have previously emphasized use of the work breakdown structure and network diagram to divide projects into small pieces of work. Each of these pieces, or tasks, has a corresponding cost estimate. In the ideal world, the person responsible for each task has prepared both the schedule and the budget estimate. This person should also have played a significant role in defining the exact Triple Constraint of his or her small work package. In any case, the project manager must assign tasks to many different people. As these tasks are assigned, some give and take in the exact scope may be accepted, but whatever is finally agreed upon must be committed to paper. That is, there should be minicontracts between the project manager and the people responsible for tasks. The minicontract defines the Triple Constraint of the task.

The project team member who now has his or her task assignment should provide the project manager with a detailed plan of how that task will be performed and periodically review progress against the plan. To the extent that the task performer has played a major role in creating and

All work assignments should be written.

initiating the task assignment, he or she is likely to be highly motivated to carry it out. Conversely, if the task was assigned without negotiation, the person may have a low sense of involvement and be largely demotivated by the assignment.

COMPUTER SOFTWARE

Figure 13-3 illustrates how Microsoft Project can facilitate team (both project and supporting) organization. Columns are created for each function (or individual) involved, and the nature of their involvement is indicated. Figure 13-3 shows marketing (Mrkt), technology (Tech), and operations (Opns) departments. Each task must have one function (or person) that is accountable (A) and may have other functions for which concurrence or input is required (C or I, respectively).

TYPICAL PROBLEMS

The usual problem is what to do with marginal personnel. This is one reason you should have inserted schedule and budget contingency—because sometimes you have personnel who must be used in an area outside their competency, which renders them temporarily marginal. In the case of truly marginal personnel, you can simply refuse to accept them on the project team. You must also expect to work with a group of people with differing and sometimes conflicting personalities. Obviously, these people must work together closely and harmoniously. Depending on the specific situation, you may have to solve individual problems, confront or replace people, or work with the entire group to improve morale and cooperation.

MATERIALS STUDY PROJECT

Imagine you are Mel Chase, SUPROMAC project manager for the materials study project I have been using as an example. Because Tom Richards is the only person in SUPROMAC capable of doing the theoretical study (tasks G, H, and J), you must have him conduct these tasks. You have two options: (1) Have

(continued)

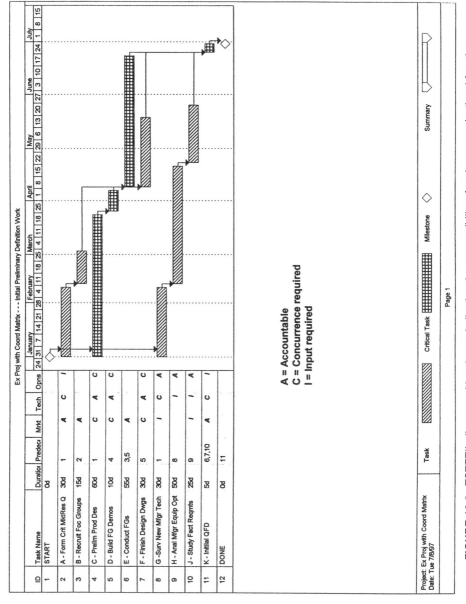

FIGURE 13-3. *TSTETIL diagram with columns indicating the responsibility of various organizational functions.*

the director of research transfer Tom Richards to your project team for the duration of the project. (2) Have Tom Richards perform the work without reassignment, as part of the support team. Thus, you would first try to persuade Richard Thomas to permit the transfer and then try to persuade Tom Richards to accept it. You could do this by stressing the importance of the project, the importance of Tom Richards's contribution to it, the reduced demand such a transfer would make on his supervisory time, and the temporary nature of the transfer. You might also stress how the experience Tom Richards would gain would benefit both him and the research section when the temporary transfer ended.

Nevertheless, such a transfer could well be undesirable for the project if you do not need Tom Richards 100 percent of each day. What would he do the rest of the time? It can also be undesirable if he is a difficult person to supervise. But if these are not obstacles, you should try to persuade Richard Thomas to approve the transfer.

Or imagine that SUPROMAC's proposal, and the resulting contract with NERESCO, designate Molly Cook (a noted mechanical expert) to work on the job, but she has just begun a long hospitalization. Your options include arranging for her to work while hospitalized, even though this may be inefficient and therefore more costly; trying to substitute a consultant or contract personnel; or trying to substitute less qualified SUPROMAC personnel.

HIGHLIGHTS

- The project team is people who work on the project and report administratively to the project manager.
- Sources of project personnel include the proposal team, others employed by the organization, and people from outside the organization.
- Compromise is required in forming project teams.
- People must join and leave the team as required during the project.
- Having team members in close proximity improves the project manager's control.

14

Organizing the Support Team

The support team is the people who work on the project either full-time or part-time for a part or all of the project but do not report administratively to the project manager. This chapter discusses how to obtain their involvement and commitment and how their efforts can and must be coordinated with the project team. I also consider interaction between the project team and support groups and subcontractors. Finally, I discuss how computer software can be used in those efforts.

INVOLVEMENT AND COMMITMENT

Involve support team members in the proposal phase or as early as possible.

As with the project team, the best way to develop a sense of involvement and obtain a commitment from the support team is to have had its members participate in the proposal. Participation also builds a team spirit that continues beyond the project. Failing this, their involvement in planning their own work and committing those plans to writing should also elicit involvement and commitment—remember the Golden Rule: Get the persons who will do the work to plan the work.

Project managers, regardless of the specific organizational form of their company, often feel they lack authority. In a project organization, project managers do have authority over the people working for them. Nevertheless, in all situations, as you will see in the next chapter, authority is not especially useful. Persuasion—especially invoking the Golden Rule—is required.

Early Support Group Involvement

Project managers and the project team often ignore support requirements, which other groups must provide, until it is too late. In larger corporations, support groups include such functions as purchasing or procurement, stores and inventory, shipping and receiving, contract administration, documentation, model shops, other specialist services, and so on. Unless support personnel understand that their services may be required, they cannot anticipate the extent to which they will be needed. Consequently, the support a project demands may not be available when needed. When support is sought tardily, support groups feel left out, and it may be difficult to obtain their commitment.

This kind of situation may arise because the project team has some degree of parochialism or is not aware what support is readily available. The project team may not understand the potential roles others can play or may assume it knows better than the support groups what kind of effort will be required. This latter situation frequently arises because the project team feels that a support group will "gold plate" the amount of work they propose to do, exceeding project budgets.

As stated earlier, these problems can best be mitigated by involving support groups in the proposal phase. If this cannot be done, involve them as early as possible in the project work. Give them an opportunity to participate in planning their task and employing their best thinking and expertise.

Have support team members estimate time and cost of their tasks.

The same applies to the time and cost estimates. The support group should make time and cost estimates for their task, and the project group should approve them. These estimates may require a negotiated revision to adjust other project tasks to accommodate support group plans if they differ from the project team's first estimate. This is a common occurrence. Support groups sometimes must perform their role at a pace dictated by other, higher priority commitments, thus scheduling your project support differently than you had planned. Sometimes the support group sees a completely different way to undertake its role, often to the project's advantage. Or the support group's experts may convince the project team that their role must be broader than originally conceived. For all these reasons, involve support groups as early as possible.

Written Commitments

Obtain meaningful commitments from support groups within your organization just as you do from the outside subcontractors, namely, a written agreement. (This is also what you should do with project team member commitments; the only difference is what actions you can take to settle disputes that may arise.) There

Put all agreements in writing.

must be a Triple Constraint and signatures by both parties. Such agreements (inside the organization) lack legal standing and enforcement provisions, but if the support group manager must sign a written agreement, he or she will be motivated to make his or her group live up to its commitment.

Support Team Advantages

As I said in the previous chapter, most project managers seem to prefer to staff their project entirely (or mostly) with project team members. However, a project manager (especially in a matrix organization) might prefer to have a large support team rather than a large project team, for the following reasons:

1. The project manager does not have to worry about the support team after the project ends.
2. In the case of subcontractors, the support agreement is embodied in a legally binding instrument, namely, the subcontract or purchase agreement.
3. The project manager has the whole world in which to find specialists or experts with the required skills.

COORDINATION

A time-based activity-on-arrow diagram aids coordination.

Once the support groups have been identified and their work has been planned properly and phased in with that of the project team, there is a continuing need to coordinate project team work. This is best done with network diagrams (Figures 14-1 and 14-2). In both figures, support group work has been segregated from the main part of the network. There are many other ways to do this, for instance, using distinctive line patterns for each support category. Where color copying machines or printers are available, a color code may be used advantageously. Some computer-based project management software allows you to generate schedules for specific groups or individuals. Other software packages allow you to dedicate portions of the schedule chart (for instance, the upper two inches, the middle three inches, or the lower two inches) to the work of a single group or individual. (Figure 13-3 shows an alternate way to facilitate coordination.)

Change

Consider the network diagram of Figure 14-2. Imagine that you discover you will be late on task D. Should you inform your subcontractor (on task B) of this lateness? It depends. In general, it is probably best to advise subcontractors of your true need date. If you do this, you make it easier for them, and their costs to you will be lower (at least in the long run). But if subcontractors have a history of lateness, it is probably best (1) to have originally allowed time contingency for their work and (2) not to let them know of any delay you have experienced.

Communication and coordination should be primarily in writing. Change should be accomplished by oral communication, over the telephone and/or at meetings involving as many people as required. But the change must then be embodied in the plan revisions.

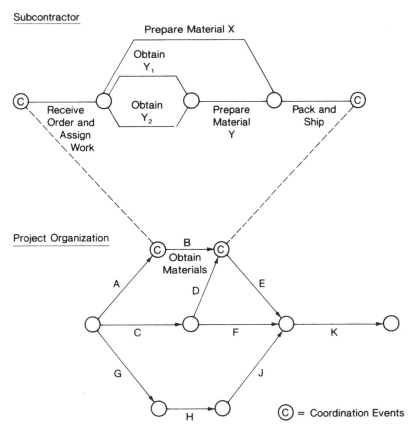

FIGURE 14-1. Network diagram illustrating use of coordination events.

Revision

Once committed to paper, plans must be disseminated to and understood by all involved personnel. Plans must also be maintained in a current status. If any out-of-date project plans are allowed to remain in circulation, the credibility of all proj-

Write and distribute plan revisions.

ect plans will become suspect. Therefore, everyone who had the original plan must receive revisions. This can be facilitated by keeping an accurate distribution list. Where your company has a local area network (LAN), each person (or at least each key person) may have a terminal on his or her desk with the current schedule. Obviously, they must both consult and understand this schedule and any changes that are inserted.

INTERACTION WITH SUPPORT GROUPS

Project team and support group interaction can be difficult. All too often the support group is brought in too late, a situation that reminds me of a story. A commuter

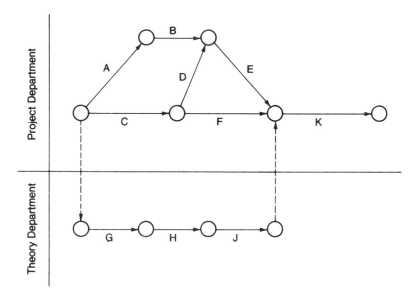

FIGURE 14-2. Network diagram illustrating use of spatial segregation of a support group's activities.

comes dashing onto the train platform just as the morning train into the city pulls out. A bystander, observing that the commuter has just missed the train, comments, "Gee, that's too bad. If you'd simply run a little bit faster, you'd have caught the train." The commuter knows, of course, that it is not a matter of having run a little bit faster but rather of having started a little bit sooner.

Project Actions

The tardy commuter's situation classically applies where the purchasing department is involved. Purchased materials arrive later than required; the project is delayed; and project people blame either the subcontractor for delivering late or the purchasing department for failing to place the order early enough. In fact, the blame lies with the people who did not requisition the purchase sufficiently early so the goods would be delivered on time. They need not run faster; they should have started earlier.

Start support group tasks early enough so that they can be finished when required.

The experienced project manager copes with this problem in two ways. First, he or she makes certain that the network diagram schedule allows enough time for the support groups to perform optimally. Second, the project manager makes certain that all personnel know when task activities must be completed and holds the task managers accountable for meeting the schedule.

Support Team Viewpoint

These issues can be looked at in a different way, namely, from the point of view of the support groups. They are composed of professionals, in the previous example, purchasing professionals who wish to obtain the best quality of required goods at the lowest possible price within the other constraints project personnel impose. They need time to perform their function in a professionally competent way. In the case of government contracts, there are even laws and regulations that require three competitive bids.

The same is true, of course, of any support group— technical writers, computer programmers, designers, draftpersons, or model shop personnel. Everyone wants to do a good job and wants sufficient time in which to do it. But departments have a workload imposed on them by others.

Support groups labor under many constraints.

They are trying to respond to many projects bringing work to them at random times and in variable amounts. Thus, support groups typically have some backlog they must work through before they can get to new requests. If they did not have this backlog, if they were sitting there idly waiting for the work to arrive, they would not be utilizing a vital organization resource, their own time, in the most effective way.

SUBCONTRACTORS

Subcontractors are basically no different than you. They have a contract from a customer, in this case, you or your project. They want to be responsive to you, but they have the same kinds of problems you do: Personnel and resources frequently are not instantly available or perfectly suitable; they need time to plan their work; they have to interpret the Triple Constraint in the correct way, and so on.

Just as a contract controls your relations with your customer, so subcontractors define their relationship to your project by the contract your company's purchasing department issues to them. Should a change be required, it is certainly all right to tell them about it. But the change becomes effective and meaningful only when it is converted into a contract change.

Another point to consider when working with subcontractors is that your request for a proposal can require that periodic reviews be included in your contract. This is desirable, as it would be if your customer required periodic reviews of your work. You are trying to see how their work is progressing, to understand if changes will be called for as a result of what they are doing or problems they are encountering, in short, to stay abreast of their work.

In many cases, you can do other things to help your subcontractors perform effectively. For instance, you may have one or more lengthy meetings with their project manager to be certain he or she knows what is important to you. You can be willing to compromise noncritical items (with suitable contract changes, if appropriate). You can understand and review their schedule to try to offer constructive suggestions. You can check analyses and witness engineering tests, provided you have made suitable arrangements.

Your support agreement is a written contract. But you must draw a fine line between giving new directions and simply keeping abreast of what they are doing. Remember, the contract dictates the work. The progress reviews or monitoring activities are not a substitute for their management of their work; rather, these are solely to find out if it is being done. You cannot provide daily, weekly, or monthly changes in their direction and expect them to be successful.

COMPUTER SOFTWARE

Figures 14-3 and 14-4 illustrate how Microsoft Project can be used to filter the TSTETIL chart previously shown in Figure 10-18 to show work by only selected resources. In Figure 14-3, the work by the engineer is shown, where one task (task E) also involves the senior engineer, who might be in a different department. Figure 14-4 shows the work of the only the researcher, from the research department.

TYPICAL PROBLEMS

Working with the support team probably causes the greatest difficulty, especially for new, inexperienced project managers. The root of this problem is being dependent on nonsubordinates. Two other problems are closely related. First, to negotiate support agreements takes a lot of time, usually at the very busy project inception period. So it is done reluctantly or poorly or even omitted. In the latter situation, the project manager uses his or her own judgment of what the support group will do. Second, even when the support agreements have been intelligently negotiated, later events frequently require that changes be made. Again, this is time-consuming and must be anticipated.

Another problem arises when the support group people are extremely busy and already have a heavy work schedule. In this situation, it is often very difficult to obtain support for the project, and the project manager lacks any direct authority. Obviously, you should try to anticipate these potential problems when the initial plans are made. You should also beware of expecting effective help from anyone assigned to work on more than two or three projects simultaneously. Effectiveness declines significantly for such persons.

Finally, resist the temptation to recruit too large a team. Larger teams impose an inefficiency "tax" by requiring more time for communication. When two firms were developing comparable mainframe processors, the smaller team achieved a faster schedule.

ID	Task Name	Duration	Start	Predeci	Resource Names
2	A - Select Materials	15d	Mon 1/1/01	1	Engineer
3	B - Obtain Materials	30d	Mon 1/22/01	2	Engineer
6	E - Conduct Experiment	55d	Mon 4/9/01	3,5	Sr. engr,Engineer

Project: Example Project Filtered
Date: Wed 8/6/97

Task ▓▓▓▓ Critical Task ▒▒▒▒ Milestone ◇

Page 1

FIGURE 14-3. TSTETIL diagram filtered to show tasks requiring the engineer.

ID	Task Name	Duration	Start	Predec	Resource Names
8	G - Review Literature	30d	Mon 1/1/01	1	Researcher
9	H - Make Theoretical Study	50d	Mon 2/12/01	8	Researcher
10	J - Write Theoretical Report	25d	Mon 4/23/01	9	Researcher

Project: Example Project
Date: Wed 8/6/97

Task

Critical Task

Milestone ◇

Page 1

FIGURE 14-3. TSTETIL diagram filtered to show work by the researcher.

194

MATERIALS STUDY PROJECT

Consider the materials study project I have been using as an example. Let us make two further assumptions about it. First, again assume that task B ("obtain materials") will require a subcontract to a material supply company. This procurement will be governed by a written contract or purchase order. Such a written instrument is valid only when it is accepted by the material supply company, usually acknowledged by the return of a signed copy. There can be no agreement unless both parties are satisfied by the contemplated Triple Constraint.

Second, assume that Tom Richards remains assigned to Richard Thomas but will be responsible for the theoretical tasks (tasks G, H, and J). All too often this kind of agreement is only verbal. But why should a $77,740 theoretical effort (see Figure 9-11) be loosely defined when a much smaller material supply effort is governed by a legally binding agreement? Support group work is commonly authorized informally, and most of the time the results are a disappointment, if not a shock, to the project team.

HIGHLIGHTS

- Support teams do not work for the project manager in an administrative sense, but their participation and contributions are vital.
- Support groups should be involved in projects as early as possible and allowed to plan their tasks.
- Although they lack legal status, written agreements are an excellent way to obtain commitments from support teams.
- Coordination, a continuing need, is best provided by network diagrams.
- Every plan revision must be written and distributed to all concerned personnel.

$$15$$

The Role of
the Project Manager

*Although the project manager is clearly involved in all phases of the project and is
ultimately responsible for satisfying the Triple Constraint, his or her interaction with the
project and support teams is a key to the leading (or "people management") phase. This
chapter first examines the overriding importance of the project manager's ability to
influence other team members. His or her leadership ability depends on motivational
skills rather than on authority, regardless of how much hierarchical supervisory authority
he or she has over project team members. Because projects are one-time undertakings,
project managers must often develop creative solutions to project problems. I suggest
ways to stimulate creativity.*

WHAT A PROJECT MANAGER DOES

Influence Rather Than Authority

As the three previous chapters note, many people working on a project do not report
directly to the project manager, and he or she does not have complete control over
even those who do. In the first place, people are free to change jobs in our society.
If given a command they do not like, some workers will simply quit. Or they may
transfer to another division of the organization. Second, modern motivational theory
indicates that issuing commands is a poor means to encourage people to perform

well on a job. McGregor's Theory X/Theory Y is but one
manifestation of the thinking that underlies current managerial practice, which usually substitutes persuasion and participation for command. Leadership is exhibited by a person
who is followed by others in the conduct of the undertaking.

**The project
manager lacks
control.**

Nevertheless, commands are still a way of life, to a greater or lesser extent,
depending upon the specific organization and situation. If stated brutally or insensi-

tively, they demotivate and create resentment. If stated politely and reasonably (which is difficult to accomplish), commands may be effective.

Given these limits to hierarchical authority, project managers must operate by winning the respect of project and support team members. This accomplished, they will find their wishes are carried out voluntarily and frequently with enthusiasm.

A study by Thamhain and Wilemon identifies nine influence bases available to project managers.

There are nine ways to have influence.

1. Authority—the legitimate hierarchical right to issue orders
2. Assignment—the project manager's perceived ability to influence a worker's later work assignments
3. Budget—the project manager's perceived ability to authorize others' use of discretionary funds
4. Promotion—the project manager's perceived ability to improve a worker's position
5. Money—the project manager's perceived ability to increase a worker's monetary remuneration
6. Penalty—the project manager's perceived ability to dispense or cause punishment
7. Work challenge—an intrinsic motivational factor capitalizing on a worker's enjoyment of doing a particular task
8. Expertise—special knowledge the project manager possesses and others deem important
9. Friendship—friendly personal relationships between the project manager and others

The first clearly depends on higher management's decision to invest the project manager with power, regardless of power's intrinsic utility. The next five may or may not be truly inherent in the project manager's position; others' perceptions are most important in establishing their utility to the project manager. The seventh is an available tool anyone may use to influence others. The project manager must earn the last two. Projects are more likely to fail when the project manager relies on authority, money, or penalty to influence people; success is correlated with the use of work challenge and expertise to influence people.

There will be occasions when the project manager must negotiate with team members. A typical approach is to explain the rationale of the effort and to involve the people in planning the detailed work packages. Given this need to influence, an effective project manager must be a superb communicator. He or she must have verbal and written fluency and be persuasive to be effective; the next chapter contains some practical tips on how to improve your communication skills.

Effective Managerial Behavior

As I said in Chapter 1, the project manager must work with people not of his or her own choosing, many of whom have different skills and interests. Furthermore, the project manager is a manager, not a doer. If the project manager is writing lines of computer code for a programming project, designing a circuit for a new product project, or nailing boards together for a construction project, who is planning the work of others? Who is deciding what approach to take to the support group manager so as to obtain the services of the most senior and best qualified person? And who is trying to devise a contingency plan in case the system test does not produce desirable results? The project manager must spend his or her time working with people and planning their work so nothing is overlooked and contingency plans are ready if needed.

A manager must plan and manage. On a very small project, the project manager's participation is also required as a worker, not merely as a manager. If not physically, then at least mentally, a project manager in this situation should have two hats, one labeled "project manager" and the other labeled "worker." The project manager must realize which function he or she is performing at any given moment and wear the appropriate hat.

Qualifications

Generally speaking, one becomes a project manager because one has been an excellent programmer, circuit designer, or carpenter rather than because one has been trained or demonstrated competency as a project manager. But a virtuoso technical performance is not a sufficient qualification for managing the efforts of the project and support teams. In fact, one's demonstrated technical or professional skills are frequently problem-solving or technical skills that do not involve an ability to interact with others. But project managers, in common with other managers, need people skills rather than technical skills (see Figure 15-1). Developing people skills can be extremely difficult for many technically trained people who become project managers. Physical systems tend to behave in repeatable and predictable ways; people do not.

The project manager must deal with many intangibles.

Project managers must be people-oriented with strong leadership and superb communication ability. They must be flexible, creative, imaginative, and adaptable to cope with a myriad of unexpected problems. They must be willing and able to take initiative with assertiveness and confidence in the face of substantial uncertainty and, in many cases, when confronted with significant interpersonal conflict. Finally, because projects are temporary, they must have a tolerance for—in a sense—putting themselves out of a job. In short, the role of project manager will be difficult for a narrow specialist (technical or other) or a person who is introverted or inarticulate.

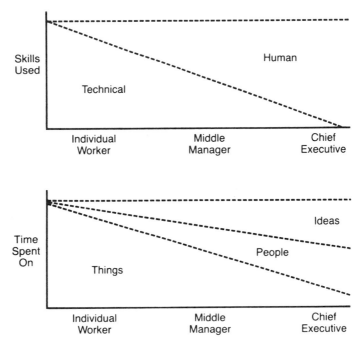

FIGURE 15-1. *Management.*

Source: "Basic Management Skills for Engineers and Scientists." Lecture notes by Milton D. Rosenau, Jr.
Copyright © 1982 by the University of Southern California. Reprinted by permission.

Working with People

Elias Porter has shown that we behave differently when everything is going well
and when we face opposition or conflict. People also differ from each other. Some
have an altruistic-nurturing orientation; other people have an assertive-directing
orientation; and others have an analytic-autonomizing orientation. Although altruistic-nurturing oriented people are usually trusting (a strength), people of another
orientation may see them as gullible (a weakness). Similarly, the assertive-directing
person's self-confidence (a strength) can be seen as arrogance (a weakness), and
the analytic-autonomizing person's caution (a strength) may appear to be suspicion
(a weakness). A person with a balance of these orientations, who is flexible, may
be seen as inconsistent.

In the case of new product development projects, the variety of required professional skills is exceptionally large. Such projects may involve marketing, research
and development scientists, engineers of all sorts, software system analysts and
programmers, industrial stylists and designers, procurement, manufacturing, and
finance personnel, and many others. Unfortunately, the training for each of these
fields tends to be exclusive, with the result that a specialist in one discipline frequently
lacks the knowledge and language fluency to understand and collaborate with people

from another discipline. One vital role of the project manager is to assure effective communications between these diverse groups.

The project manager should be chosen because of an interest and skill in human relations.

To compound this problem, the same thing said to the same person at two different times can produce different reactions. This lack of predictability can be a major pitfall for many prospective project managers. Project managers must deal with both technical and emotional issues. If not already fluent with these human relations skills, they should take a course in behavioral psychology.

Managers manage and workers perform the tasks.

The project manager sets objectives and establishes plans, organizes, staffs, sets up controls, issues directives, spends time working with widely varied people, and generally sees the project is completed in a satisfactory way, on time, and within budget. The project manager does not do the work of others on the project. A project manager who is an excellent electronic engineer may find it frustrating to watch a junior engineer carry out the circuit design activities on the project. The junior engineer will take longer, make mistakes, and not do as good a job as could a project manager with that technical skill. But if the project manager starts to do the circuit design, it demotivates the junior engineer and lessens the manager's time to function in the most vital role of all, namely, that of project manager.

THEORIES OF MOTIVATION

Regardless of hierarchical authority, anyone can encourage or stimulate others to contribute to a project and improve their productivity. For project managers, many of whom lack direct authority over all the resources required for project success, an understanding of motivation is essential. A major element of the project manager role is to avoid demotivating others.

Consider the following problem, which a project manager called to my attention:

> If I fail to meet a schedule I prepared myself, I feel obliged to work overtime (frequently unpaid) to try to meet the estimate; however, if the schedule or estimate was prepared by someone else, I generally feel under no obligation to try to meet it. That is, the work takes whatever it takes. I'm not sure this is a reasonable attitude. Nevertheless, regardless of what is logical, it is the way I actually feel that controls my motivation.

Another project manager described the situation of a co-worker who had been severely criticized by his boss. As a result, that co-worker is no longer motivated and is both argumentative and defensive about many things he is now asked to do. In both of these situations, motivation—or the lack of it—is a challenge for the project manager. Thus, an understanding of motivational theory is important. Although the basic work is fairly old, it remains important for you to understand. By analogy, we still teach Euclidian geometry and Einsteinian relativity even though they were formulated a long time ago. There are many theories of motivation, the two most important of which I briefly review.

Hierarchy of Needs

Abraham Maslow's theory of the hierarchy of needs (see Figure 15-2) holds (and human experience validates) that motivation is not external but rather arises within the worker if managerial actions are not inappropriate. The worker is motivated to achieve a specific goal because of an inherent internal need. Each person has five levels of needs. The first

People are motivated to fill their unsatisfied needs.

of these is physiological or body needs (eat, sleep, have shelter, and so forth). A hungry person will have a goal to eat and will engage in the goal-directed activity of buying and preparing food to satisfy the hunger need. (In fact, if we try to prevent this person from satisfying the need, by denying him or her money to pay for the food, for instance, the person may engage in such antisocial behavior as robbery to satisfy the need.)

Once this need is filled, continuing to offer more food, sleep, or shelter has no motivational value. Higher level needs now come into play. Second-level needs are safety and security; third-level needs are social; and fourth-level

People have five levels of needs.

needs are esteem or ego. The fifth-level needs are for self-actualization or fulfillment. To take advantage of Maslow's findings, a project manager would have to understand the levels of need an individual has already satisfied. Then he or she could offer

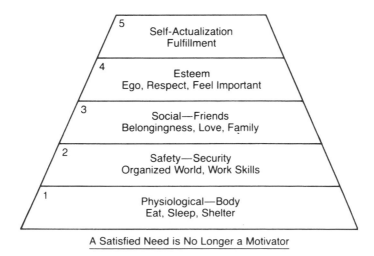

FIGURE 15-2. *Motivation—Abraham Maslow's hierarchy of needs.*

TABLE 15-1   Satisfiers

Level	Job Related	Other
5	Setting own goals	Hobbies or volunteer work
4	Winning company award	Election to civic board
3	Staff meetings or committee work	Club membership
2	Insurance packages	Live in safe area
1	Enough salary to purchase necessities and reasonable work hours	Availability of adequate stores

Source: "Basic Management Skills for Engineers and Scientists." Lectures notes by Milton D. Rosenau, Jr. Copyright © 1982 by the University of Southern California. Reprinted by permission.

satisfiers for unmet needs as an encouragement. Americans generally have unfilled needs on levels three, four, or five; therefore, the project manager would offer satisfiers aimed at these levels.

If the project manager does not offer satisfiers, workers will find them outside the work environment (see Table 15-1). Note that money is not a motivator, although it may play a role. A person feels good when he or she gets a salary increase because it provides recognition (filling an ego need). However, if a person can fill social needs only by joining a country club, he or she needs more money than the person who can satisfy these social needs with informal activities.

Finally, you must realize that there are some tasks on a project for which it may be impossible or exceptionally difficult to motivate a worker. One project manager asked me, "How do you motivate people to follow through on the less glamorous aspects of the project, such as punch list items (that is, minor residual details)?" He could have asked about the necessity to clean up laboratory equipment or other similar tasks. The best suggestion is to try to make people realize that all work is a mix of the mundane and the exciting and that they should harvest some of the euphoric "high" they get from the stimulating tasks to carry them through the depressing "low" of the necessary but boring ones.

Motivational Factors

Frederick Herzberg has done an excellent study on work and motivation, examining specific factors that motivate workers. He found many things done (company policies, supervision, work conditions, and salary, for instance) that are not motivational at all. He called these "hygiene factors." The absence of hygiene factors is demotivating, but their presence is not motivating.

Achievement and recognition are the most powerful motivators.

Instead, there have to be motivational factors present to foster high productivity. The key motivational factors are achievement, recognition, the work itself, responsibility, advancement, and growth. Achievement and recognition are short term, and the others are long term in their impact. Thus, it is important to give workers recognition frequently (but not routinely) for significant accomplishments.

IMPLICATIONS OF MOTIVATIONAL THEORY

How can you take advantage of these widely accepted theories of how people become motivated? There are many techniques, three of which I briefly review.

Theory X/Theory Y

I have already mentioned Douglas McGregor's work in identifying two managerial styles, Theory X and Theory Y. Theory X is the authoritarian style, in which top management makes decisions and coerces workers to comply. Theory Y is the participative style, built upon the findings that people both enjoy working and want to work. Theory X assumes external control of the workers; Theory Y assumes useful controls are within individuals and managers can draw upon workers' self-direction.

In working with the support team, over which he or she lacks control, the engineering project manager should try to adopt the Theory Y style (and would probably be well advised to use it with the project team as well). Authoritarian behavior is not usually appropriate. Managerial behavior can be based on the Theory Y assumption that others want to do a good job.

Participative decision making is best.

To put it another way, consider how you would like to be managed and how you would respond. Workers are rarely self-motivated in a Theory X environment, but are more likely to be so in a Theory Y environment.

Behavior Modification

B. F. Skinner devised a theory of behavior modification that advocates positive reinforcement, namely, rewards for "good" behavior. Using Skinner's theory, one would induce people to behave differently (that is, consonant with project goals) by rewarding them when they act appropriately. "Rewards" typically are consistent with Herzberg's findings, namely, a sense of achievement and recognition.

Thus, if a designer does a fine job on your project, it is appropriate to send her a memo and send a copy to her boss (or vice versa) and perhaps to her personnel folder as well. Conversely, if you have been practicing behavior modification and positive reinforcement consistently, you do not have to do anything if on another occasion she does a poor job. The absence of positive reinforcement will be message enough. Further, that absence may very well motivate the designer to ask you how her performance fell short.

Negative reinforcement can stop undesired behavior, but it requires positive reinforcement to promote desired behavior.

Very early in my industrial career, the president of my company sent the following letter (on engraved, personal stationery) to me at my home just after the project team of which I was a member shipped the first unit of an advanced system:

Dear Milt:

I would like to extend to you my hearty congratulations for your contribution to the outstanding technical success of the Satrack Program. I know that you must feel proud of being a member of the team who accomplished a marked advance in the state of the art of aspheric manufacture.

Our Company is now considered to be in the forefront of this development activity, and it has been through your contributions that we have achieved this position.

I know that your efforts were great and there were many long evenings and weekends which you personally sacrificed. It is indeed gratifying that we have the people with the spirit to undertake such a challenging problem and carry forth to a successful conclusion.

Sincerely yours,

People's behavior can be modified. The recipient of such a letter will become motivated to put in similar extraordinary efforts in the future and probably will continue to have family support for that effort. Other examples of positive reinforcement include awards (wall plaques, luncheons or dinners, trips, and so forth) or a story or picture in the company newspaper. These promote worker motivation because they provide recognition and validate an achievement.

Management by Objectives

Another practical technique that the project manager can use is management by objectives (MBO). Variations of this technique are management by results (MBR) and management by commitment (MBC). As the names suggest, the techniques focus management and worker attention on the outcome (objectives or results) rather than on the process by which a worker chooses to achieve the outcome. This frees the manager to concentrate on what is desired, and it leaves the worker free to concentrate on how to accomplish it in his or her own style.

To use this simple, powerful technique, the project manager negotiates with a worker on the results the worker will agree to achieve. Because the technique is a general one, the worker may be a member of the project team or the support team. The agreed results must be like project objectives, that is, verifiable, measurable, specific, and achievable. These are recorded on paper (perhaps a standard form) and are signed by both the worker and the manager. In the event that changes are **MBO is a powerful, useful technique.** required, these must be jointly negotiated, and the paper must then be revised and signed again. Note that this technique is identical to how you would contract with another organization to do subcontract work; the only difference is that the MBO agreements are not legally binding.

The technique draws upon the motivational factors of responsibility and achievement. The worker may be motivated because he or she is consulted in a responsible way in defining the task to be carried out. And there may be clear progress during the work that provides the worker with a periodic sense of achievement.

The only problem with the technique is that it is sometimes difficult to set the objectives or results that are to be achieved. When this is the case, the same technique used for starting a project with unclear objectives can be used. That is, agree on something specific that will be accomplished in a given period, and then try to set longer term goals when more is known about the task. Conversely, if you cannot establish a specific goal for a worker to achieve, it is clearly unfair to hold him or her accountable for something you could not describe.

STIMULATING CREATIVITY

Frequently, the project manager must also stimulate creativity. This is required to some extent on all projects, but it is especially applicable to high-technology projects. Even two similar projects, for instance, building two identical radio telescopes, can encounter different soil conditions requiring an innovative approach in one case. In general, the less precedent for the project, the more creativity will be required.

Motivation

Creativity may be stimulated simply by managing in a way consistent with behavioral and motivational theory. Encouragement by providing recognition and appreciation is the most straightforward technique to stimulate creativity. In addition, one must provide a favorable atmosphere. In a sense, we are looking at positive reinforcement again. People are permitted to fail when asked to produce creative results and are not castigated for doing so. Rather, they are praised when they succeed.

Brainstorming

Brainstorming techniques are often used to deal with some intractable problems. The conventional method for brainstorming is to advise perhaps a half-dozen people of the problem and after one or two days convene a brainstorm meeting. At this meeting, restate the problem and reiterate the ground rules:

Positive reinforcement and brainstorming will stimulate creativity.

1. Absolutely no criticism (including smirking) is permitted.
2. The more ideas produced, the better is the session.
3. Novel, unusual (even impractical) ideas are desired.
4. Improvement or combination of prior ideas is also desirable.

Use a tape recorder to permit more leisurely subsequent consideration of the ideas thus generated.

    An alternative method that works better in most situations is to have a facilitator talk to a few people individually and ask for their ideas on solving the problem. After three or four people have been interviewed, the facilitator will have a list of ideas to use to start the brainstorming session (like

Facilitated brainstorming is very productive.

pump priming), which is then carried out in the normal way. This facilitated brain-storming produces more ideas and these are of better quality.

TYPICAL PROBLEMS

> Project managers are not normally selected from a pool of trained, qualified people. Rather, projects arise within (or descend upon) the organization, and a person who has demonstrated technical proficiency is asked to become project leader.
>
> Such people are often good "doers" and have technical skills and may think they want to be a project manager, but they usually take the job not knowing what is involved. In general, technical specialists rapidly master planning techniques and then the mechanics of project monitoring. They may get along with others (as opposed to being hermitlike) but be unable to cope with the inevitable conflicts that bedevil the project manager. Or they may be poor communicators. What then happens is the organization has a poor project manager and has lost the services of a good technical specialist.
>
> One cure for this problem is to be sure that candidates for project management read books such as this prior to being offered jobs as project managers. After that, assuming a continuing interest in the job, the selected candidates should be offered further training, as discussed in Chapter 27.

MATERIALS STUDY PROJECT

> At SUPROMAC, Ed Ross has been concerned for some time about the develop-ment of more project managers and has periodically conducted in-house seminars on project management. Thus, Mel Chase, who has demonstrated competency as a mechanical engineer as well as exhibiting drive and personal ambition, attended such a seminar two years before he was asked to manage the NERESCO materials study project.
>
> However, in common with most other engineers, Mel has still not mastered many of the engineering project management skills. Thus, he remains unsure of himself in working with people. In such a situation, both he and his management must seek out further training to improve this most critical skill. Because this is often less threatening when done without the inhibition of co-workers, Mel locates a university seminar that deals with his areas of concern. He requests authorization to attend, and Ed Ross concurs and thus approves attendance at SUPROMAC's expense.

HIGHLIGHTS

- There are many limits to authority; so project managers should learn how to wield influence.
- Managers must confine themselves to planning and let others perform the tasks.
- Human relations skills are vital to a project manager.
- Familiarity with theories of motivation will help managers do their job.
- Creativity can be stimulated by positive reinforcement and brainstorming.

16

Practical Tips for Project Managers

Because a project manager must be a superb communicator, I first discuss the general problem of communication and then provide several simple suggestions for improving your communication skills. Another pervasive problem for project managers is resolving conflicts, and some techniques to deal with conflict are provided. I also briefly discuss how a manager can gain time to work more effectively with people. Finally, I offer some miscellaneous pointers.

COMMUNICATION

Communication must be worked at.

Effective communication is one of the more difficult human endeavors. There are so many obstacles it is amazing that any effective communication at all occurs. Words have different meanings, and people often have different perceptions or orientations. The project manager's reputation (be it as a jokester or as a very serious person) will alter the way any message is received. Everyone the project manager communicates with will tend to hear the message he or she wants or expects to hear, which is not necessarily the message the project manager is attempting to deliver. Sometimes people are not listening, are distracted, or have a closed mind.

There is an aphorism about how to communicate: First, you tell people what you intend to tell them; then you tell them; and then you tell them you told them. There is much truth in this use of multiple message delivery.

There are several general steps you can take to improve your communication with other people:

Plan what is to be communicated beforehand rather than trying to decide while communicating. As it is sometimes stated, "Put brain in gear before opening mouth."

Use face-to-face meetings in which you can observe the other person's "body language." Allow enough time at an appropriate time of the day.

Decide which sequence and combination of telephone discussion, face-to-face meeting, and memo will be most effective.

Use E-mail carefully and thoughtfully. Much E-mail is quite casual, is sent hastily or carelessly, is given insufficient attention by recipients, and can cause unnecessary confusion. E-mail does have the advantage of quickly reaching many people at dispersed locations if they are connected "on-line." Of course, they must have time to read, digest, and act on this torrent of messages.

Be consistent and follow through with actions appropriate to your message.

Use simple language.

In addition, you should consider using feedback, notices, and proximity.

Feedback

Communication is very much like a servomechanism in that it is not effective unless there is feedback. Communication can be improved by asking the person to whom the message has been delivered to restate it in his or her own words. This can help overcome a listener's closed mind. Another

Effective communication requires feedback.

effective technique is to back up any verbal communication with a memo. This may also be done the other way around, first sending the memo and then having a meeting to discuss it. The duality of mode and the recipient's restatement, rather than being simple redundancy, is most effective here.

Notices

It is impractical to meet constantly with all participants on a very large project. Even on a smaller project, it may be disruptive to have numerous meetings. It is thus desirable to issue project notices and reminders of priority actions for any given period. Putting such notices on distinctively colored paper or preprinting the project name on the top will set them apart from other mail. In addition, electronic mail and other forms of modern telecommunications can be used. If you are using project management software with a local area network, you can deliver many notices through that software.

Proximity

Locating the people on the project near each other also aids communication. Because the people are close together, they can see each other more often, which makes communication easier and more frequent. And when people are in frequent contact, their points of view tend to become more uniform.

Follow-Up

It is necessary to have some system of follow-up of the communications, be they face to face or written. Some people simply keep an action log, a chronological listing of all agreements reached with other people for which follow-up action is expected.

Somewhat more effective is a follow-up system keyed to the individual from whom action is expected. A filing card with each key person's name printed on the top may be used to record notations of actions expected of that person. A variant of this is keeping a folder for each key person in which you store records of all discussions or copies of memos for which follow-up action is required or requested. In either event, hold periodic meetings with each key person and use the filing card or folder to plan the topics to be discussed.

Follow up communications.

When it is known that project managers (or any manager, for that matter) have such a consistent follow-up system, people who work for or with them will realize that any statements made to them will be taken seriously. Therefore, commitments made to them tend to receive serious and consistent attention.

CONFLICT RESOLUTION

Projects are fraught with conflicts. They inevitably arise because projects are temporary entities within more permanent organizations. One root cause is thus competition for resources. Regardless of organizational form, project managers and functional managers tend to have momentary interests that are at odds, so project managers must expect and be able to "stomach" conflict. If you have a low tolerance for conflict, being a project manager can be frustrating.

The project manager must cope with conflict.

Studies have shown three ways to reduce conflict:

Constantly anticipate and try to reduce conflict.

1. Work at conflict reduction proactively because conflict is always present and will not go away if you ignore it or wish it were not present. In fact, left unconfronted, it normally gets worse.
2. Have and maintain good plans in the form of current and realistic schedules agreed to by the people involved.
3. Communicate thoroughly with all involved people and their management.

EFFICIENT TIME MANAGEMENT

Given the wide range of project managers' duties (in a sense, they must be all things to all people), they can easily end up working nights and weekends unless they are very efficient in the use of time. Of course, they should not make the mistake of being efficient to the point of being ineffective. Effectiveness is achieving the desired results. Effectiveness is what counts, but project managers are more likely to be effective if they use their time efficiently.

Manage time efficiently.

The overriding issue in time management is "first things first." The project manager must know the most important things to do this year, this month, this week, today, and right now. One project manager told me that she spent two-thirds of her available time in meetings, focus group testing panels, training seminars, and serving on committees and that only about one-eighth of that (two-thirds) time had anything to do with the new product development projects for which she was responsible. Only when he or she has a clear perception of priorities can a project manager effectively manage his or her time. Thus, each day select a few (not more than three or four) really important, high impact matters for your primary attention.

The second key issue in time management is to devote large chunks of time to important single issues (for instance, one two-hour meeting with a colleague, rather than twelve fifteen-minute meetings). This can be accomplished by maintaining a time log on how you will actually spend your time next week (See Figure 16-1). At the end of the week, examine the record of what actually occurred, as you recorded it at the end of each time interval during the week. Then plan how you will alter your behavior in the following week to get fewer, larger chunks of time concentrated on single topics. Do this several times over a period of several months and you will master the art of better time management.

Because project management involves integrating the work of many people, numerous meetings will be held. As some anonymous sage has stated, "Meetings are an institution where the minutes are kept and the hours are thrown away." Conducting them efficiently and effectively is essential. The following are keys to improving meetings:

Know beforehand why the meeting is to be held and what outcome is expected.
Consider whether it is possible to omit the meeting.

FIGURE 16-1. Sample time log.

Source: "Basic Management Skills for Engineers and Scientists." Lecture notes by Milton D. Rosenau, Jr. Copyright © 1982 by the University of Southern California. Reprinted by permission

Determine the minimum number of people required.

Choose a meeting location with a room arrangement consonant with the meeting's purpose (for example, a round table arrangement for discussion among equals, a lecture hall arrangement for a presentation, and so forth).

Plan all meetings and always circulate an agenda in advance.

Circulate an agenda with topic durations to all attendees, and perhaps discuss this individually with the key participants ahead of time.

Be prepared, and open the meeting on time with a re-statement of the purpose and agenda.

If possible, ask each attendee (one at a time) for his or her views on each topic prior to topic completion. Just because some attendees are loud or dominant does not mean they are best qualified to speak on a topic; a quiet, shy, or retiring person will often make a valuable contribution if invited to comment.

Verbally summarize what transpires at the meeting and later distribute published minutes to all attendees.

They won't work everywhere, but you might want to try the meeting rules on the Gossamer Albatross project (as reported in *Technology Review,* April 1981, p. 56):

All meetings are held standing in a circle.

All participants are heard in turn.

All meetings must result in a definite decision.

All decisions must be acted on immediately.

PRACTICAL TIPS

There are a few "tricks of the trade" that project managers should employ to help put the techniques discussed in this chapter to practical use.

First, keep your door open. This encourages people to talk with you, and sometimes that will identify key project issues that you were not yet aware of.

Second, close your door and do not answer the telephone. This is when you do your planning, to gain high leverage on time use. When you are also a worker on your project, which is a common situation on smaller projects, this is when you do your own project work. In addition, a closed door permits private discussions if these are required. Finally, a closed door is a trivia filter, which may force people to stand on their own feet and make minor decisions.

Third, walk the halls. There are always some people who will not enter your office, even if the door is open. Also, what you inevitably see when you go to the sites where work is supposed to be done is that things are not as you expected them to be. At Hewlett-Packard, this is called management by walking around (MBWA).

Fourth, set a good example. Arrive early (or at least punctually) for work;

take your job seriously; work hard; and be respectful of others (especially upper management and your customer), even if you disagree with some of their actions.

Fifth, when you are encountering problems, do not try to conceal them. Rather, seek advice from more senior people.

Sixth, remember that one of the most effective things a project manager can do is change resource allocations. Rather than pushing harder on (or shouting at or abandoning) a task manager who is experiencing difficulty, try to get him or her additional help.

TYPICAL PROBLEMS

> The development of your own natural style of management may be a problem when applying the practical tips in this chapter. My style of dealing with conflict or improving time management, for instance, may not be best for you. You must find your own way to deal with limited time and multiple demands on it, and you must recognize that interpersonal conflicts can be very distracting and impede productivity.
>
> Another problem is balancing the two roles of project manager and worker, if the project is such that you must play both roles.
>
> In general, however, the biggest problem is to become a superb, versatile communicator. You must work with varied people—many (or most or even all) not of your own choosing—and that can be done only if you communicate effectively with them. Some of the people you must deal with may themselves be poor communicators, as either message deliverers or receivers. Some may receive and reply to written memos, and perhaps you can write these well. Or (and this will always be the case with some workers) they may not read or understand written memos. So you must talk with and get useful responses from these other workers. Obviously, the reverse may be true also, that is, your writing skill may be relatively weaker than your speaking skill. But some workers receive written information better than spoken; so you may have to improve your writing (or reading) skills.

MATERIALS STUDY PROJECT

> Mel Chase walks around to each group working on the materials study project every week, spending a full hour with each responsible task manager. (Because there are never more than three tasks running in parallel, this requires only a morning or afternoon each week.) He varies the day and time from week to week to avoid predictability. If his walk around time were predictable, some task managers might always try to avoid him, and some others might try to "dress up" the appearance of their work.

HIGHLIGHTS

- Effective communication can be aided by feedback, issuing notices, and locating workers near each other.
- Conflict between people (or their group managers) must be expected on any project. Conflict can be reduced by having plans that are current and by good communications.
- A project manager can be more effective if he or she is a good time manager.

Part 4

Monitoring Progress on a Project

17

Monitoring Tools

The next managerial activity on a project is monitoring progress. First this chapter discusses various monitoring techniques. Then there is a detailed consideration of the use of reports. Next I talk about using computer software to monitor project progress. Finally, I discuss the special case of monitoring several projects simultaneously.

CONTROLLING TO ACHIEVE OBJECTIVES

The word "control" has a pejorative connotation, implying power, domination, or authority. Thus, many project managers (especially new ones) tend to avoid the necessity of installing and using controls on projects. The management of the $2.2 billion project to automate many of the Immigration and Naturalization Service's functions was criticized in early 1997 because, in part, it lacked clear milestones that would clarify whether it was on schedule and within budget. The purpose of such project controls is to measure or monitor progress toward your objectives, evaluate what needs to be done to reach these objectives, and then take corrective actions to achieve the objectives. Thus, you must employ controls (in the measurement sense of the word) or your project will go off course and you might never know it.

Controls are needed to monitor actual progress compared to the project plan.

CONTROL TECHNIQUES

The first, and in many ways the most important, control is a well-publicized plan for all three dimensions of the Triple Constraint. A work breakdown structure, a network diagram that indicates every element of the WBS, and a cost estimate for each activity indicate how the project should be carried out. Any deviation—and there normally are several—from this three-dimensional plan indicates the need for corrective action. Without such a plan, control is impossible.

What you measure is that to which attention will be given.

There are several restrictive control tools available, such as withholding resources or discretionary authority. When the project manager uses these controls, he or she is assured that people working on the project request the use of these resources or authorities, thus providing visibility. As an example, the project manager could require any expenditure in excess of $1,000 to receive his or her specific approval. Or the project manager could require that any drawing release have his or her signature. These kinds of controls go beyond the project plan in that they make project workers seek out the project manager for approval during the performance of each project activity or task. Anyone's failure to request approval of a planned major purchase tells the project manager that the project has deviated from the plan.

Controls can tell you if the project plan is being followed.

This kind of restrictive control may well be appropriate with an inexperienced team or on a difficult project. But it is normally appropriate only for very small projects. If essentially all decisions on a large project must flow through the manager, the project will get bogged down by his or her lack of time to review a myriad of documents for approval. An effective variation on this restrictive approval control approach is to insist on independent inspection and quality control approvals or on test data as means to verify progress. For instance, you could insist that each subsystem test be approved by people working on other related subsystems.

Another project control method is to place trust entirely in the person carrying out a particular task. This is a form of management by exception, and what you want to have brought to your attention is any "political" problem or deviation from performance, schedule, or budget. This method is fine if that person is able to recognize deviations from plan and realizes they must be reported promptly to the project manager. The person must also be capable of reporting the problem clearly. Because these three preconditions are rarely satisfied, this control tool should not normally be used.

Control is best exercised by examining the status of tasks.

A far better approach is for the project manager to examine the work being done under the direct control of the project team and support teams. This kind of control is based on the Theory Y assumption that people working on project tasks will be trying to do a good job (which can often become a self-fulfilling prophecy), rather than on the Theory X assumption that people will not do a good job. These examinations of activity often may be accomplished by reading reports and conducting project reviews.

REPORTS

Reports fall into three broad categories: those concerned with the accomplishments along the performance axis, those concerned with schedule progress, and those concerned with cost. Reports may be written as summaries to provide an overview or be detailed about a particular task activity or some other element of project work. Reports may be strictly for the use of the organization performing the project or be

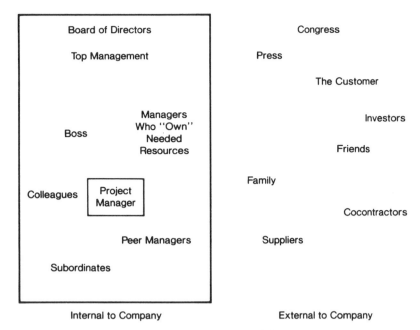

FIGURE 17-1. *Report recipients.*

intended for people outside, such as the customer or contractors. If the project organization is an industrial company, such as SUPROMAC, there is a wide spectrum of people to whom reports might be addressed, as shown in Figure 17-1.

Problems

The project manager cannot depend entirely on reports. In the first place, they may be inaccurate, a common failing of cost reports, which are prone to arithmetical errors and data entry errors if a computer-based system is being employed. On the other two axes, the people who write reports

Do not confuse activity with progress.

are prone to unwarranted optimism. Project workers generally assume a well-advanced task is nearly complete; in fact, no task is complete until it is truly finished. Thus, most task reports will indicate a task is 80 or 90 percent complete, implying it will require additionally only a small fraction of the time already spent. The following news report about the Stamford (Connecticut) railroad station illustrates that percent complete can be a very misleading measure of project completion status:

> With a 15-ton sculpture planned for the new railroad station here, a city engineer tested the floor beams to make sure they could support such monumental artwork.
>
> Afterward, the engineer revealed "some bad news and some worse news."

"If the sculpture is installed," he said, "the floor will cave in."

"And the worse news?" asked (the) mayor.

"If the sculpture is not installed," the engineer replied, "the floor will still cave in...."

When work stopped, the station was considered 85 percent finished. But now, officials say, much of the project must be demolished and rebuilt.

Copyright © 1985 by The New York Times Company. Reprinted by permission.

Unless the task is as simple as drilling a hundred holes in a steel plate, it is normally impossible to measure what percentage of the task is truly complete. (And, as I said before, even in this case, there is no assurance the last hole won't destroy the plate, forcing the task to be completely redone.) In general, completing any given task requires more time than the person working on its forecasts. For instance, a report that X engineering drawings have been completed for an engineering project originally planned to require Y engineering drawings tells you nothing about the percentage of completion. You know you are truly complete only when Z engineering drawings (which may be either more or less than the Y drawings planned) actually are completed, checked, and released. The same is true for lines of code if the project is software and, by analogy, will apply to all other types of projects as well.

Using 80 percent complete as a specific example, there are two very different ways that percent complete might be reported:

Eighty percent of the tasks are 100 percent complete (which assumes that all tasks are known and approximately equal).

One hundred percent of the tasks are 80 percent complete (which assumes that the remainder of the work on tasks is equal to the completed work and that no other tasks will be required).

In most cases, when people report percent complete, they do not identify which of these two bases they are using. In general, percent complete is reported based on the amount of time or budget consumed (or an average of these). Percent complete is used frequently in large contract projects as the basis for progress payments. There are handbooks for the construction industry that provide guidelines on how much erected steel or poured concrete is worth, for example; but as we saw in the case of the Stamford railroad station, there is no assurance that the complete work is actually satisfactory.

Percent complete is fundamentally flawed for progress monitoring.

Even though the task manager's report expresses a high degree of confidence that the task will be completed when scheduled, such a promise is not deliverable. Reports are one-way communication and lack the give and take possible in a meeting; so it is hard to judge the status of a task in progress solely from written data.

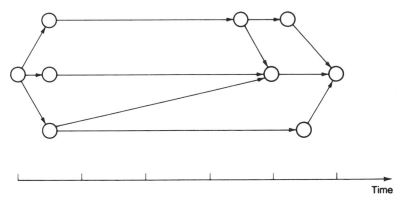

FIGURE 17-2. Time-based activity-on-arrow diagram.

Consider the schedule situation illustrated in Figure 17-2. There is a long period of time in the middle of the project when there are no scheduled task completions. Thus, there are no certain checkpoints available, leaving considerable uncertainty as to the actual status. This illustrates another reason to break a project into many small tasks because there will be less uncertainty as to schedule status.

Concentrate on completed task deliverables, not activities.

If the work breakdown structure divides the project into many small tasks, the manager can look at each of them individually and decide whether they are complete. Only those tasks for which all work has been done are considered complete. No other task, regardless of the amount of effort applied so far, is complete. A task that is 80, 90, or 99 percent complete is not complete. This two-state (bivalued) approach to examining tasks simplifies the project manager's job tremendously. He or she can accept as complete only those tasks for which reports (oral or written) guarantee the activity is complete. In fact, the task authorization document (Figure 6-4) could have additional blocks for the signatures of the task manager and project manager to indicate their agreement that the task is satisfactorily complete (see Figure 17-3). With many small tasks rather than only a few larger ones, there is much less uncertainty about overall project status. There are other benefits of using smaller tasks:

Dividing a project into many small tasks aids precise determination of its status.

Inherently less risk
Improve goal clarity
Faster reinforcement upon completion
Greater sense of urgency because the deadline is sooner
Earlier warning of difficulty

TASK AUTHORIZATION		PAGE OF
TITLE		

PROJECT NO.	TASK NO.	DATE ISSUED

STATEMENT OF WORK:	COMPLETION REPORT:
APPLICABLE DOCUMENTS:	

SCHEDULE START DATE:	COMPLETION DATE:
COST: PLANNED =	ACTUAL =
ORIGINATED BY: DATE:	ACCEPTED BY: DATE:
APPROVED BY: DATE:	APPROVED BY: DATE:
APPROVED BY: DATE:	APPROVED BY: DATE:
COMPLETED BY TASK LEADER:	ACCEPTED BY PROJECT MANAGER:

FIGURE 17-3. *Task authorization form modified to serve also as a task completion report and acknowledgment form.*

Detail Level

Reports should always be as brief as possible. Because many people concerned with the project will wish to receive reports, there is a tendency to try to circulate one report to many recipients. This is a mistake. Many who wish to be kept abreast of progress are not interested in small details. This is especially true of busy management, both project organization and customer, for whom the project manager should prepare brief, summary level reports. Such reports might also be circulated to people concerned with some specific aspects of the project but who do not require all the details in any particular report.

Pictures, demonstrations, and models should also be encouraged, especially for a high-technology project or one where much of the work is geographically remote. It is often hard for people not intimately concerned with the project to visualize the status, expected outcome, or even the concept. For them, tangible descriptions, pictures, and such are by far the most appropriate means for providing reports. If your organization has closed-circuit television, this may be an excellent use of it. Some companies transmit computer data by satellite or other relay, permitting a geographically remote work site to construct a model or visualize design information in a three-dimensional display. The ubiquitous fax machine also allows people at other locations to see sketches and pictures in close to real time.

It is always important to avoid the pitfalls of excessive reporting. Clearly, this is a gray area requiring judgment. The "convenience" of copying machines can easily lead to too many copies of overly detailed reports being circulated to too many people.

Finding simple, highly visible measures can be helpful. Figure 17-4 shows one way to report schedule variance from plan. The number of days until completion is periodically graphed, and any deviation from the "no slippage" line (drawn at 45 degrees) reveals slippage. (A horizontal line shows no progress.) Figure 17-4 has a line above the "no slippage" line indicating a small slip, which might be reasonable. Another simple device is to arrange for calendar clocks that count backward to the scheduled completion.

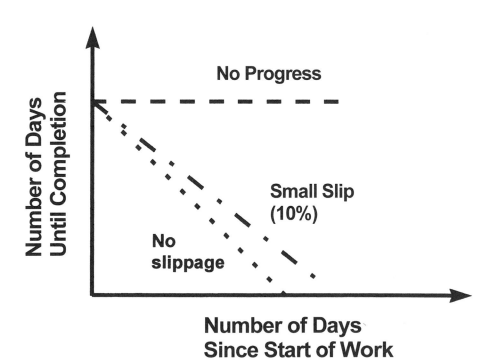

FIGURE 17-4. *A simple schedule report to quickly reveal slippage.*

Cost Reports

Neither the passage of time nor the expenditure of money is a measure of progress.

Cost planning and reporting are important because they provide a guide for management action. They permit comparison of actual accomplishments in terms of cost incurred and planned accomplishments. Thus, they aid in determining whether there will be a cost overrun or underrun at completion. In some instances, they may aid in determining whether a specific task is on schedule.

Unfortunately, in many organizations, there is a tendency to look at actual cost versus planned cost as virtually the only measure of project progress. This is useless because there may be offsetting overruns and underruns, which I discuss in the next chapter.

MULTIPLE PROJECTS

As a project manager, you may eventually be responsible for more than one project. You may have many project managers reporting to you, or there may be so many task managers reporting to you on a single project that you cannot personally attend all the task reviews and critically examine all the necessary detail. In this case, you must receive some kind of summary information that indicates the status of the several projects (or the many tasks) for which you have responsibility.

The value of summary status indicators is questionable.

When you cannot personally get into details, you are completely at the mercy of those who summarize information for you. In the early 1990s, the secretary of defense, a man ultimately responsible for many large developmental projects, had to apologize to Congress (of which he had previously been a member) for misinforming them of the status of a navy stealth attack plane. He had not been fully or accurately informed of the project status (late and overrun) by many of his aides, several of whom then resigned or were reassigned. The project was then canceled. In this case, the advantages of managing multiple projects (you can average your successes and failures and you may have flexibility to move resources from one project to another) were outweighed by the disadvantages (dependence on summary information, coping with simultaneous problems, and having only limited time to devote to a given project).

Personal reviews are best.

Probably a better approach is to visit the various project reviews personally. Imagine you have three projects reporting to you, each of which is being managed by another project manager, who reviews it monthly. You might sit in on each of these in a rotating fashion; so you attend the review of each project once every three months. If one project you have responsibility for is significantly more important than the others, attend these reviews more frequently (for example, every other month, if not monthly).

COMPUTER SOFTWARE

Figure 17-5 is an example of how you might replan the example project shown in Figure 7-19 to make it easier to monitor. In this illustration, two critical path tasks (C and E) that were long (sixty and fifty-five days, respectively) have been subdivided into three shorter subtasks.

TYPICAL PROBLEMS

Problems arise because of too much or too little monitoring. The former may demotivate personnel, consume too much time, or cost too much. Inadequate monitoring, based perhaps on the naive optimism that the project will be performed in accordance with plan, can lead to disaster. The following (edited) quotation comes from a memo written months after the start of a new product development project:

A. The accessories were far more complex than anticipated at the time the original plan was prepared. There was not, at that time, any defined level of performance for these accessories; the feasibility of the accessories had not been studied. The estimates for the optical systems and mechanical components of the accessories were inadequate until several months of feasibility study had been performed and the products fully defined. Even after they were defined, all of the accessories have had to go through repeated design iterations. In October, all of the accessory designs were rejected. The Head Adapter and the Lamp Housing required mechanical design rework to reduce their size. In addition, the Lamp Housing required optical redesign to accommodate additional optical features necessary for a smaller, more compact package. The Head Adapter was finally approved in mid-December; however, the base was not approved until mid-January. The Lamp Housing received verbal approval only last week. The Accessory, as designed in this project, was found to be completely unacceptable, requiring total redesign, forcing its removal from this project (all future efforts on the Accessory will be covered by a separate project and are excluded from this summary).

The original plan of last February (13 months ago) contained approximately 600 hours of optical and mechanical design for the Accessories. In our revised plan, including labor to date, the optical and mechanical design is approximately 2400 hours; the difference is 1800 hours plus drafting time.

B. The original plan called for 670 hours for industrial design. Because the designs have been subjected to repeated revisions, the revised plan, including labor hours spent to date, puts industrial design at approximately 1700 hours: an increase of 1030 hours.

(continued)

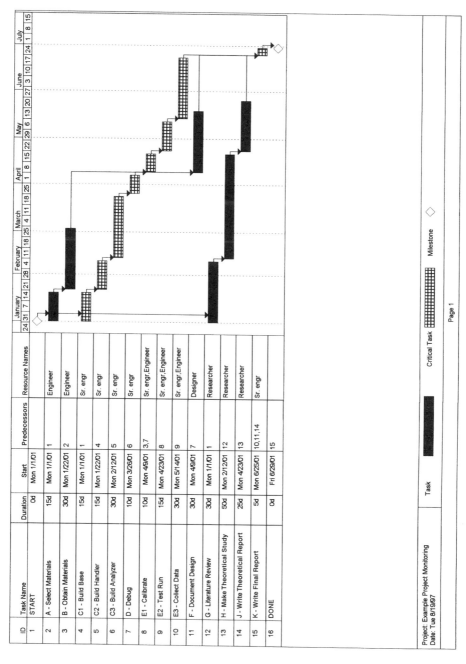

FIGURE 17-5. *TSTETIL schedule using smaller subtasks on the critical path to improve monitoring.*

C. The layout and design of the Indicator Dial required considerably more time than that provided in the original plan. The Dial design was rejected several times, which added approximately 800 hours.

D. The custom microcircuit for the digital meter version added coordination time with the vendors and with purchasing of approximately 300 hours.

E. It was assumed in the original plan that the proprietary microcircuit would be available as an input to this project. Delays in receiving properly functioning devices made the timing critical. Therefore, coordination with the vendor and evaluation of sample devices of about 400 hours was performed on this project.

F. Due to the complexity of the mechanical design, we added preparation of a tooling plan, preliminary tool design, and coordination to the project. This has added approximately 450 hours. (We will also do the tool design, but that work is covered under a separate project.)

G. The revised plan contains 90 hours for planning and supervision; that time is now charged to the project rather than handled as indirect overhead.

H. The original plan was based on burdened labor rates that were in effect when the plan was prepared. The change in burdened labor rates since then has added approximately 20 percent to the project cost.

I. Personnel being shifted temporarily to other projects has caused approximately 200–300 hours to be added.

J. Approximately 340 hours were added to support engineering time to better match experience on other projects.

K. The original plan allowed for about 800 hours for project administration. The actual and planned hours now represent an increase of approximately 800 hours due to slightly higher monthly hours plus the longer duration of the project.

L. Compared to the original plan, a higher level of technical documentation hours have been expended. Additional drafting time associated with the Accessory mechanical designs and redesigns (due to size rejections) is approximately 600 hours. Other factors were the low estimates for standard parts drawings and reworking circuit layouts due to mechanical and size limitations. These other factors represent approximately 1200 hours.

This project has gotten into severe cost (approximately quadrupled) and schedule difficulties, and it basically got that way by slipping one day at a time. Continuous monitoring could have detected deviations from plan in time to redirect—or terminate—the project.

In addition, there is a monitoring problem because cost (and other) report time lags delay news of project difficulties. In some instances, task leaders or other key people may not be motivated to submit progress reports. This problem is not uncommon where these people are technical experts who are more comfort-

(continued)

able with their technology than with writing. Further, there are inaccurate reports, even from conscientious people. In other cases, unclear reports will mask the deviation that has actually occurred.

Also, even if the reports are prompt, accurate, and clear, the deviation may be noticed when you are busy with other urgent activities. Whenever a deviation is noticed, it takes time to react. Finally, you may be distracted by a human tendency to search for a guilty party on whom to blame everything.

TASK AUTHORIZATION	PAGE *1* OF *1*

TITLE *Select Materials*	

PROJECT NO. **83627**	TASK NO. **83627. 01**	DATE ISSUED **3 JAN**

STATEMENT OF WORK:	COMPLETION REPORT:
Select specific test materials that conform to our proposal.	*Have chosen X and Y as meeting requirements of NERESCO, and satisfying our need for price and availability.*
APPLICABLE DOCUMENTS: *Proposal (attached)*	

SCHEDULE	
START DATE: **3 JAN**	COMPLETION DATE: **20 JAN**
COST PLANNED = **$5,250. —**	ACTUAL – $ **$6,850. —**
ORIGINATED BY: DATE:	ACCEPTED BY: DATE:
APPROVED BY: DATE:	APPROVED BY: DATE:
APPROVED BY: DATE:	APPROVED BY: DATE:
COMPLETED BY TASK LEADER:	ACCEPTED BY PROJECT MANAGER

FIGURE 17-6. Task completion report for task A of materials study project.

MATERIALS STUDY PROJECT

Assume that Mel Chase has to decide how to control the materials selection and procurement tasks (tasks A and B) of the materials study project. Because much of the work will be routine, he opts to trust the task manager, Cheryl Williams. A discussion with Cheryl has made Mel confident that the three critical conditions are satisfied: She can recognize if there is a problem. She knows a deviation must be reported promptly. She both writes and speaks clearly. As further assurance and control, Mel asks Cheryl also to provide a brief written note on work status each month.

At the conclusion of three weeks, Cheryl reports that she has concluded task A on schedule, using the task authorization (Figure 17-6). Assuming Mel Chase agrees that the task is complete, he would also sign the form and make copies for the project file and any other people who have to be notified.

HIGHLIGHTS

- Comparison with the project plan provides the basis for monitoring a project.
- Monitoring by noting completed tasks provides an unambiguous measure of progress.
- Projects managers may exercise control by requiring their approval or trusting task managers, but the best approach is to examine the status of tasks.
- Reports, which may concern any axis of the Triple Constraint and be detailed or general, provide a means to examine the status of tasks.
- Managers in charge of several projects can best exercise control by periodic personal reviews.

18

Project Reviews

There are two kinds of project reviews; periodic (typically monthly) and topical. Reviews are a very important control tool available to the project manager; they assure that he or she and the workers will actually be meeting to discuss progress. This chapter deals with both types of reviews and the general necessity to conduct reviews.

THE NECESSITY FOR REVIEWS

Having reviews is very much like having a navigator on an airplane. Reviews and a navigator are unnecessary if everything is proceeding according to plan. The purpose of both is to uncover deviations and correct them. Experienced project managers know the project will not proceed as planned, but they do not know how it will deviate. Only the naive project manager believes the plan is sufficient and no further navigation is necessary to arrive at the project's Triple Constraint point destination.

Reviews are your off-course alarm.

The project manager's boss and other senior management will frequently want to know about the project status. Although this may not be true for relatively insignificant projects within an organization, it is a very common situation for commercial development projects where company money is at stake and for larger engineering or construction projects being performed on contract for others. The customer (in external projects), too, may wish to have periodic or topical reviews. These requirements and their nature and thoroughness should have been included in the negotiated contract for the work. The people working on the project will also wish to have reviews of the overall project from time to time. This is their means of learning whether their effort has to be adjusted from plan to conform with some new reality or everything is still proceeding according to the original plan (which never occurs). Reviews with the project personnel in attendance are a means for communication and can enhance their motivation.

THE CONDUCT OF REVIEWS

Whether reviews are periodic or topical, they should be planned. Certain questions are almost always appropriate to raise.

Planning

Reviews may be thought of as a very small project. The goal is acquisition of all relevant information. There is a schedule and cost. The schedule may be a simple statement that the review will consume two hours on a particular

Think of a review as a small project.

afternoon. The cost depends on the number of people who participate and the preparation time.

There should be a plan for reviews. Everyone involved should understand the Triple Constraint for the review and be prepared to carry out their assignments. This means the project manager must know how much time and what level of detail are appropriate for their participation. Reviews may also be thought of as a particular kind of meeting. As such, all the care of preparation and follow-up discussed in Chapter 16 is relevant.

Questions

The smart project manager learns to ask questions at project reviews. You are not asking questions to embarrass or pillory anyone, but rather to find out how the project is deviating from the plan so you can take corrective action. You want

Ask nondirective questions.

to learn about what has and has not happened so far, but you can change only the future. Although questions about the past can help you anticipate what must be done in the coming days or weeks, they may be intimidating if something has gone wrong. Thus, ask questions nonthreateningly. If you (or others such as configuration management or quality assurance) don't ask questions, some people won't volunteer critical information. Examples of nondirective questions include the following:

What is your greatest concern?

Do you anticipate any problems that we haven't talked about yet?

What persistent problems do you have, and what is being done to correct them?

Do you need any resources (people or things) you do not yet have?

Are there any personnel problems now or that you anticipate?

Do you know of any things that will give you schedule difficulties in completing your task? If so, what are they?

What kind of help would increase your confidence in the schedule?

What schedule risks have you considered?

Is there anything I can do to help?

Is there any possibility that completion of your task will lead to any technological breakthroughs for which patents might be appropriate?

Has any work done on your task led to any competitive edge we might use to gain other business elsewhere?

There are judgmental or backward looking questions that may be harmful:

What went wrong?

Who made that mistake?

Why didn't you anticipate that?

In most cases, these areas should be explored in private sessions and even there the questions can be reworded to be less threatening. For example, the first question might be restated more usefully as "What would you do differently next time?"

> **Plan to conduct project review meetings and expect problems to surface.**

The thing to remember about project reviews and these questions is that you are almost assuredly going to hear some bad news. Most of us do not cope with bad news in a very positive way; so the project review can easily become a recrimination and blaming session. This will not be productive. It will destroy the review and much additional effort on the project. Be businesslike and factual in conducting the reviews, and keep asking questions to gather information.

PERIODIC REVIEWS

In general, every project should be reviewed once a month. It may be appropriate to review some projects once a year or once every three months; others may require weekly or even daily reviews. Periodic reviews can catch deviations from plan before they become major disasters. In the case of the overrun new product development project discussed in the previous chapter, periodic project reviews could have caught the deviations at the end of one or two months, when something might have been done about them, rather than at the end of thirteen months, when the accumulated deviation from the plan was so severe.

Task Review

A task is in one of two conditions: complete or not complete. For tasks whose performance axis dimension has been completed, examine the actual versus planned cost and schedule, as illustrated in Figure 18-1. Assuming all tasks were estimated on a uniform basis, for instance, a 50:50 likelihood of underrun or overrun, and unless there is something unique about the cost deviations on any completed activity, the accumulated actual cost versus plan can be used to project the cost at the end of the project. In Figure 18-1, actual cost for the five complete tasks ($42,000) is

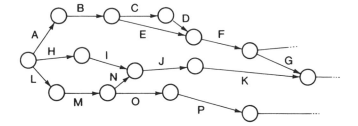

1 - Identify technically complete activities.
2 - Examine actual versus planned cost and schedule:

Completed Activity	Cost			Schedule		
	Actual	Plan	Variance	Actual	Plan	Variance
A	3,000	3,200	200	7	6	<1>
B	4,100	2,900	<1,200>	9	5	<4>
C	6,000	8,400	2,400	6	11	5
D	9,700	12,600	2,900	12	17	5
E	19,200	17,000	<2,200>	24	23	<1>

FIGURE 18-1. *Measuring progress.*

less than planned cost ($44,100), and the ratio of these indicates that the final actual costs will be approximately 97.5 percent of the plan.

Actual versus plan ratio may not be meaningful in the case of schedule variations because many of the completed tasks will not be on the critical path. An activity not on the critical path will often be completed later than plan simply because it was not necessary to complete it within the *Watch the schedule and recalculate the critical path.* planned time. Thus, the project manager can make predictions about the schedule only by looking at completed tasks on the critical path and then redetermining it. It is also important to watch near-critical paths because these sometimes become critical. As I will show later in this chapter, project management software is quite helpful for this.

It is appropriate to ask which incomplete tasks are in progress and which have not yet been started. For those under way, find out whether there have already been any difficulties that would preclude their being completed on time within the cost plan.

In the case of the project illustrated in Figure 18-1, the next concern would be the status of tasks H, L, and F. Do these tasks not yet completed indicate that the project is hopelessly behind schedule? After exploring that, we want to know about the critical path for this particular project. In this case, the concern is whether the completed activities have already caused the project to slip or whether the schedule variations have no significance with regard to the overall project.

In the case of cost review, it is necessary to examine the details of each task individually, as shown in Figure 18-2. The project manager who looks only at the overall project may be deceived. Project cost can appear to be in harmony with the plan, but that may conceal compensating task overruns and underruns or other difficulties that must be explored. Naive managers assume that actual cost in excess of plan is bad (task A, Figure 18-2), but that may only signify work being completed early. Similarly, actual cost underrunning plan is naively assumed to be good news (task B, Figure 18-2), but this may actually be symptomatic of a situation in which actual progress is inadequate. People may be avoiding the task because it is difficult or unpleasant. To determine, the real situation, the project manager must ask questions, not just look at project cost estimates.

Commitments

Project cost reports are always plagued by certain problems. First, they are not issued instantly at the end of the month. It takes time to gather the data, process it, print it out, and distribute it, whether the system is manual or electronic. Second, these reports are never the highest priority in any corporation's accounting department, coming after payroll, reports mandated by law, and usually after customer billing. Thus, project cost reports typically follow completion of the monthly period by two or three weeks (at least a week and a half, and, in some cases, five or six weeks).

Commitments are future costs.

In addition, cost reports are nothing but a record of the apparent charges to a project, which may be in error. Even if the cost charges to the project are correct, they do not necessarily reflect the corresponding amount of performance

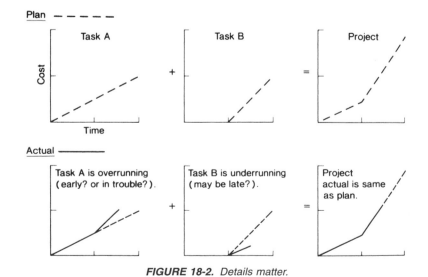

FIGURE 18-2. *Details matter.*

Month	Event	Commitment Report	Cost Report
1	$10K purchase order issued	$10,000	-0-
2	$1K trip authorized	11,000	-0-
3	Traveler paid $900	10,000	$900
4	Vendor delivers	10,000	900
5	Vendor sends bill	10,000	900
6		10,000	900
7	Vendor's bill paid	-0-	10,900

FIGURE 18-3. *Commitments are future costs.*

accomplishment. If the actual costs for each task agree completely with the plan for each task, there is no assurance that the work accomplishment has a corresponding degree of completion. Within an individual task, where there is no detailed further subtask breakdown, the project manager has only judgment to guide him or her in determining whether progress and cost are consistent. But if a project is divided into many tasks, the ambiguity is reduced because the project manager can look at completed tasks and compare their actual cost with plan. In looking at these costs for tasks, the manager must also look at commitments. Commitments are obligations for which the project will be billed but that are not reflected on the actual cost reports. Figure 18-3 illustrates two common problems.

A task apparently completed with an underrun may in fact have outstanding bills charged against it. To preclude surprises, look at the commitments charged against the task. You may do so with either a manual or computer-based commitment report. If these reports do not exist, the project manager must maintain these manually to avoid unpleasant surprises.

To summarize, review the Triple Constraint:

- Technical dimension—Task is either complete or not.
- Schedule dimension—Compare actual and planned time for completed tasks. Slippage is significant only if the project's schedule is delayed.
- Cost dimension—Compare actual and planned cost for completed tasks. Include commitments.

Cost to Complete

Periodically, the project manager should request estimates of the cost (and time) to complete all incomplete tasks. You might do so quarterly or semiannually and also whenever the plan is revised substantially.

	Method 1		Method 2	
Plan (or Allowed) Cost	=	$10,000	Task Q	=	$1,632
Cost to Date	=	6,000	Task R	=	2,149
Cost to Complete	=	$4,000	Task S	=	1,910
			Cost to Complete	=	$5,691

FIGURE 18-4. *Estimated cost to completion.*

Do not make these estimates by simple subtraction of cost to date from the planned cost, as in method 1 in Figure 18-4. Have each task manager totally reestimate the task in light of everything now known, as shown by method 2 in the figure. Unfortunately, this often indicates that the project will overrun its budget, but it is important to learn of that possibility early enough to do something about it. As the project progresses, the participants learn about what is involved. Also, replanning may occur, and the necessity for new tasks may become apparent.

FOLLOW-UP ACTIONS

———
Follow-up reviews.
———

During any review, a variety of actions will be identified to cope with the various problems uncovered. The project manager should always record these actions, the person responsible, and the expected completion date. This might be done on a register (Figure 18-5). All concerned people and their supervisors should receive copies, and the status of these action assignments should be reviewed no later than the next project review.

TOPICAL REVIEWS

There are many kinds of topical reviews for projects. The one used depends on the undertaking and customer requirements. Many new product development, engineering, and other hardware projects have the following types of reviews:

1. Conceptual design
2. Preliminary design
3. Critical design
4. Manufacturing
5. Preshipment

```
┌─────────────────────────────────────────────────────┐
│                                                       │
│            Assigned Actions for Completion            │
│                                                       │
│               Project: _____               │
│                                                       │
├──────────────────┬──────────────────┬────────────────┤
│      Action      │   Responsibility  │      Due       │
├──────────────────┼──────────────────┼────────────────┤
│                  │                  │                │
│                  │                  │                │
│                  │                  │                │
│                  │                  │                │
│                  │                  │                │
│                  │                  │                │
│                  │                  │                │
│                  │                  │                │
│                  │                  │                │
│                  │                  │                │
│                  │                  │                │
│                  │                  │                │
│                  │                  │                │
└──────────────────┴──────────────────┴────────────────┘
```

FIGURE 18-5. *Typical action follow-up form.*

6. Management

7. Customer

The exact titles for these topical reviews will vary from organization to organization and will differ somewhat for software and construction projects.

All these reviews are often mandated in the contract and may be a precondition to carry out further work on the project. In such a case, the reviews themselves would clearly be designated as task activities on the network diagram and be considered major project milestones, as discussed in Chapter 7.

Management and customer reviews are often onerous, but they may stimulate participation and involvement on the part of all people working on the project. For this to happen, the project manager must solicit ideas from all project personnel as to what should be discussed during the review. Rough ideas for the review should be delineated in a smaller group of key staff. At this point, the manager should delegate portions of the review to other people. (Be sure there are not so many people making presentations at the formal review that it becomes a circus.) Next conduct a trial run with a fairly large group of project participants invited to criticize.

Following this, prepare materials for the formal presentation, before which it is desirable to conduct a second dry run with a peer management group that represents the same range of skill backgrounds as the audience that will attend the formal review. The people attending this dry run review will provide additional insights as to how materials can be better presented or changed. If possible, videotape the presentation for later review.

Major review presentations can be an opportunity.

Now you are ready to conduct the formal review, after which it is desirable to conduct the entire review again for the benefit of everyone working on the project. They are just as interested as you in the kinds of questions that management and the customer asked and will find it just as interesting and motivating to spend some time hearing what happens and getting an overview of the particular issues covered in the review.

COMPUTER SOFTWARE

Project management software is particularly valuable for highlighting the impact of schedule changes when the actual time to complete a task is different than the plan time. Figure 18-6 shows that there is a co-critical path when the literature study task (task G) requires ten weeks. Figure 18-7 shows that the lower branch of the network (containing the theory tasks) has replaced the previous critical path if the literature study task requires twelve weeks.

TYPICAL PROBLEMS

Reviews are plagued by three common problems. First, there is always a concern as to whether the information being presented or discussed is accurate. Cost reports, as previously noted, are especially prone to error. Beyond this, there is often speculation about the exact status of some task, component, or report. Clearly, good planning for reviews can reduce if not eliminate this problem.

The second problem is the poorly conducted review. Aimless discussion or recriminations are common. Running a review like any other well-planned meeting greatly reduces the possibility of getting off the track.

The third problem is that some project personnel fail to meet schedule deadlines and do not consider this to be a major problem. For example, technologists may believe it is more important to arrive at a perfect or elegant solution (however long it takes) rather than an adequate solution (on schedule). "Better" is the enemy of "good enough."

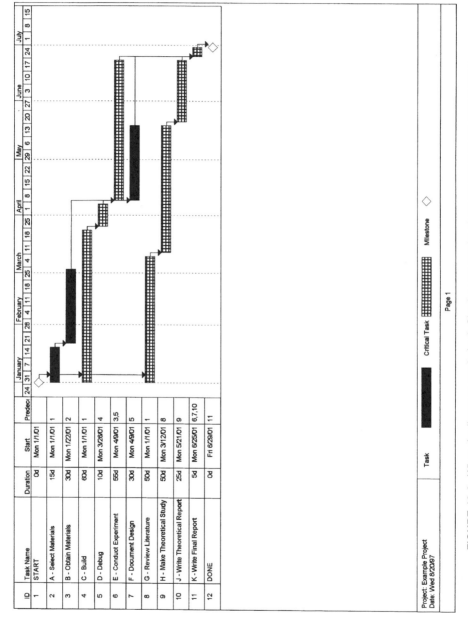

ID	Task Name	Duration	Start	Predec
1	START	0d	Mon 1/1/01	
2	A - Select Materials	15d	Mon 1/1/01	1
3	B - Obtain Materials	30d	Mon 1/22/01	2
4	C - Build	60d	Mon 1/1/01	1
5	D - Debug	10d	Mon 3/26/01	4
6	E - Conduct Experiment	55d	Mon 4/9/01	3,5
7	F - Document Design	30d	Mon 4/9/01	5
8	G - Review Literature	50d	Mon 1/1/01	1
9	H - Make Theoretical Study	50d	Mon 3/12/01	8
10	J - Write Theoretical Report	25d	Mon 5/21/01	9
11	K - Write Final Report	5d	Mon 6/25/01	6,7,10
12	DONE	0d	Fri 6/29/01	11

Project: Example Project
Date: Wed 8/20/97

Task ▬ Critical Task ▦ Milestone ◇

Page 1

FIGURE 18-6. When the literature study (task G) takes ten weeks, there are two parallel critical paths.

239

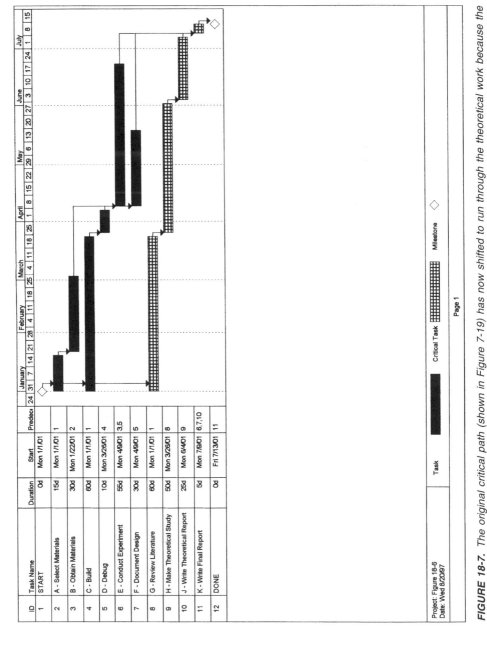

ID	Task Name	Duration	Start	Predec
1	START	0d	Mon 1/1/01	
2	A - Select Materials	15d	Mon 1/1/01	1
3	B - Obtain Materials	30d	Mon 1/22/01	2
4	C - Build	60d	Mon 1/1/01	1
5	D - Debug	10d	Mon 3/26/01	4
6	E - Conduct Experiment	55d	Mon 4/9/01	3,5
7	F - Document Design	30d	Mon 4/9/01	5
8	G - Review Literature	60d	Mon 1/1/01	1
9	H - Make Theoretical Study	50d	Mon 3/26/01	8
10	J - Write Theoretical Report	25d	Mon 6/4/01	9
11	K - Write Final Report	5d	Mon 7/9/01	6,7,10
12	DONE	0d	Fri 7/13/01	11

Project: Figure 18-6
Date: Wed 8/20/97

Task ▨   Critical Task ▨   Milestone ◇

Page 1

FIGURE 18-7. The original critical path (shown in Figure 7-19) has now shifted to run through the theoretical work because the literature study (task G) has come to require twelve weeks.

240

TASK	COST ($)			SCHEDULE (WEEKS)		
	Plan	Actual	Variance	Plan	Actual	Variance
A *	5,250	6,850	<1,600>	3	3	—
B *	9,125	8,725	400	6	8	<2>
▷ C *	87,270	89,890	<2,620>	11	11	—
▷ D *	10,900	10,900	—	2	2	—
▷ E	177,200	P		12	P	
F *	92,400	81,200	11,200	6	5	1
G *	18,220	19,300	<1,080>	5	10	<5>
H	42,020	P		11	P	
J	17,500	N		5	N	
▷ K	14,950	N		1	N	
TOTALS FOR COMPLETED TASKS	223,165	216,865	6,300			

Legend

▷ Critical Path
* Task Complete
P Task in Progress
N Task not Started

FIGURE 18-8. Summary of material study project after four months. Cost projection is for 3 percent underrun, but critical path has now changed, and project will be one week late unless something changes.

MATERIALS STUDY PROJECT

At the fourth monthly project reviews, the summary information in Figure 18-8 is presented to Mel Chase. All tasks that are supposed to be completed have actually been completed, as has Task F, and progress is being made on the

(*continued*)

two other tasks. In addition, actual costs to date for completed tasks are below plan. But, although there has been no schedule slippage of tasks on the critical path, a careful examination of schedule shows that the original critical path has been replaced by a new critical path (the rest of task H, plus tasks J and K), due to the very late completion of task G.

Consider another possibility. Imagine that SUPROMAC management requires a preshipment review before the final report (task K) can be sent to NERESCO. Because this is not shown explicitly as a separate task, it must be integral in the task (that is, comparable to typing, editing, binding, and so forth). Thus, when the project enters the last month, Mel Chase might prudently begin to schedule the appropriate managers' time to participate in the review.

HIGHLIGHTS

- Reviews uncover the inevitable deviation from plan and allow a consensus as to the needed corrective action.
- Reviews, like projects, must be planned.
- Ask nondirective questions, and expect problems to surface at project review meetings.
- Periodic reviews should be conducted as appropriate for the project, but once a month is a good rule of thumb.
- The kind of topical review used depends on the project and the customer's requirements.
- Questionable accuracy and poor procedures are common problems with reviews.

19

Project Cost Reports

A project cost accounting system is required to monitor the cost dimension. Such monitoring may reveal schedule or performance dimension problems. I discuss these topics and include examples of cost reports, which are normally generated as computer printout. I also discuss a hybrid system, Cost/Schedule Control System Criteria, used in large Department of Defense projects.

COST MONITORING

The availability of information to monitor a project's actual cost varies between organizations. Any firm that performs cost reimbursable projects under contract to another organization must have a project cost accounting system that can be audited. Commercial customers may not choose to conduct such an audit, especially if the contract value is relatively small, but governmental organizations routinely conduct audits during and after projects. Governmental organizations may also audit project costs on fixed price projects. Thus, firms doing work for the federal and other governmental entities and many firms in the construction industry have well-developed project cost accounting systems that capture labor and nonlabor project costs.

Firms must also treat capital financial investments differently than costs or expenses that can be written off in the current financial period. Therefore, when a firm carries out a project that adds to its capital infrastructure, it must collect and report those costs. A typical situation is the addition of a new production line in a factory, where plant engineering labor costs and other direct expenses are capitalized along with the purchase price of the newly installed production equipment. Such capital costs may be audited by tax authorities. Although the firm's system to collect and record such capital costs may not be as elaborate as the one for contract projects, a system must exist.

Other firms carrying out projects may not have systematic project cost accounting systems. One example is many firms conducting new product development projects.

The important issue in such firms is before-the-fact planning of what resources will be required, the time to market, the expected profit to be earned, and the payoff in comparison to alternative investments in other new products. Once the project is under way, costs per se are of relatively little importance compared to staying on schedule. All costs (except any capital expansion to build new or specialized production capacity) are expensed in the current year, and these are typically controlled by limiting the staff headcount level.

Sunk costs do not justify continuing a project.   Many projects, especially new product development projects, are terminated prior to plan or market entry (or later, before the original profit goals are achieved). Sometimes project managers mistakenly argue to continue a project because a large investment has already been made. This is an error because expensed costs are merely written off (although unrecoverable capital costs must be treated differently). Throwing "good money after bad" is never justification to continue a marginal project.

COMPUTER COST REPORTS

Many organizations, especially large ones and those performing projects under contract to a governmental entity, commonly have computer-generated reports that summarize project cost. Such reports normally come from mainframe or minicomputers. In some cases, they may also cover schedule deviations. There is a variety of report systems available for purchase or lease. Figure 19-1 shows how such reports can be useful for project monitoring. Actual cost (and schedule) data are collected from labor time cards, purchase orders, and other direct charges to the project. These are compared with plans; noted variances are analyzed; and required corrective actions may then lead to replanning. Comparing reality to the plan may suggest that certain trends will lead to future variances, which again is a cause for replanning. To be useful, this comparison must be done for each cost center (for example, department) and work breakdown task.

Variance from plan is a danger signal.   The project manager must steer a careful line between having too many forms and too much information and having too little information to monitor the project, although it is probably better to have more information than less. Neither the project manager nor top management should be trapped into believing that actual cost data (which can be reported with great precision) are the only measure of project health (which requires difficult three-dimensional measurements).

Most organizations with computer support for project management issue a weekly labor distribution report. This report lists the names of people charging time to each project during the previous week. This key report for project managers provides an early warning signal. Examination of the report may reveal that people are charging your project who should not be charging it and that people whom you expected to be working on it are not charging your project.

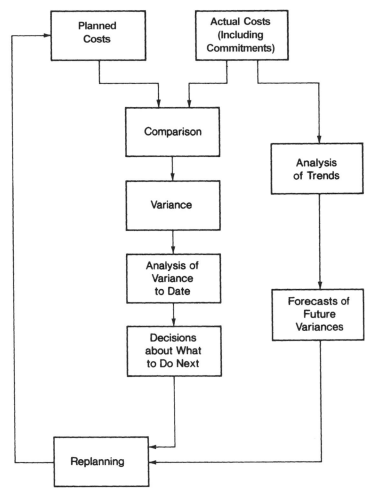

FIGURE 19-1. *Replanning is called for whenever future trends or past actuals indicate significant deviation from plan.*

COST MONITORING PROBLEMS

This section reviews a few of the cost reports for the materials study project, illustrating some issues in project control. Actual report formats vary in different organizations, and the level of detail can range from minimal to extensive. (Figure 19-2 is one example of the kind of information that would be provided by a reasonably helpful cost reporting system.) These reports are typically prepared with a mainframe computer, working with the planning data base previously discussed. But it is not absolutely necessary to use a computer, although the amount of data to be handled can otherwise be a substantial burden, even to a well-qualified project cost accountant.

PROJECT COST REPORT

PROJECT_____MATERIALS STUDY_____ TASK__B-OBTAIN MATERIALS__ DEPARTMENT__CHEMICAL__

CATEGORY		MONTH 1 - JAN			TOTALS TO DATE		
		PLAN	ACTUAL	VARIANCE	PLAN	ACTUAL	VARIANCE
HOURS	SR. PROF.	40	40	0	40	40	0
	JR. PROF.						
	SR. TECH.						
	JR. TECH.						
DOLLARS	LABOR	1,000	1,000	0	1,000	1,000	0
	OVERHEAD	1,000	1,000	0	1,000	1,000	0
	NONLABOR	4,400	0	4,400	4,400	0	4,400
	PRIME COSTS	6,400	2,000	4,400	6,400	2,000	4,400
	G & A	960	300	660	960	300	660
	TOTAL COSTS	7,360	2,300	5,060	7,360	2,300	5,060

FIGURE 19-2. *Cost report for first month for task B work by the chemical group.*

Variances Due to Timing

Figure 19-2 is the project cost report for task B within the chemical group at the end of the first month. Typically, this report would be available about the middle of the second month (February, in this case). In this case, labor and the overhead thereon are in accordance with plan. But there is a favorable variance in the non-labor expense. That is, there was a plan to spend $4,400, but nothing has been spent.

A variance between actual cost and plan may be due to payments being made later than plan rather than to work variances.

Figure 19-3 shows the corresponding task and period commitment report for the chemical group. Commitments in the amount of $4,400 have been incurred. Thus, the cost variance merely indicates that bills have not yet been paid rather than being a variance due to activities not yet undertaken. To put it another way, it is anticipated that the $4,400 expense will occur later. This demonstrates that it is impossible to make intelligent use of project cost reports without also examining commitment reports.

There may also be variances caused by purchasing a large amount of some material, some of which is needed for an early task and the rest for a later task (for instance, to obtain a volume discount). In this situation, the receipt (and cost) of material do not conform to task usage.

Figures 19-4 and 19-5 are the same reports for the end of the second month. Once again, the labor hours and costs as well as the overhead are in accordance

PROJECT COMMITMENT REPORT

MONTH ____1____

PROJECT <u>Materials Study</u> TASK <u>Obtain Materials</u> DEPARTMENT <u>Chemical</u>

Commitment Date	Item	Amount	Estimated Payment Date
21 Jan 92	P.O.—Material Supply Co.	$4,000.	30 Apr 92
21 Jan 92	Travel Auth.—C. Williams	400.	15 Feb 92
		$4,400.	

FIGURE 19-3. *Commitment report for first month for task B work by the chemical group.*

with the plan. In this case, there is an unfavorable variance during the second month with regard to nonlabor costs because the travel voucher is paid now but the plan had the expense in the first month. For the totals to date, that is, through the end of the second month, the nonlabor variance is favorable. This favorable variance is composed of the unpaid $4,000 purchase order, which is variance only because of payment timing, and a $50 favorable variance because the travel voucher payment was $50 less than plan.

Variances Due to Actual Work Not As Per Plan

Figures 19-6 and 19-7 are the project cost reports at the end of the third and fourth months for the same task. In this case, labor hours exceed plan in the third month, with

PROJECT COST REPORT

PROJECT____MATERIALS STUDY____ TASK__B-OBTAIN MATERIALS__ DEPARTMENT__CHEMICAL__

CATEGORY		MONTH 2 - FEB			TOTALS TO DATE		
		PLAN	ACTUAL	VARIANCE	PLAN	ACTUAL	VARIANCE
HOURS	SR. PROF.	4	4	0	44	44	0
	JR. PROF.						
	SR. TECH.						
	JR. TECH.						
DOLLARS	LABOR	100	100	0	1,100	1,100	0
	OVERHEAD	100	100	0	1,100	1,100	0
	NONLABOR	0	350	(350)	4,400	350	4,050
	PRIME COSTS	200	550	(350)	6,600	2,550	4,050
	G & A	30	83	(53)	990	383	607
	TOTAL COSTS	230	633	(403)	7,590	2,933	4,657

FIGURE 19-4. *Cost report for second month for task B work by the chemical group.*

PROJECT COMMITMENT REPORT

MONTH ____2____

PROJECT Materials Study TASK Obtain Materials DEPARTMENT Chemical

Commitment Date	Item	Amount	Estimated Payment Date
21 Jan 92	P.O.—Material Supply Co.	$4,000.	30 Apr 92
21 Jan 92	Travel Auth.—C. Williams	All	Paid
		$4.000.	

FIGURE 19-5. *Commitment report for second month for task B work by the chemical group.*

attendant unfavorable cost variances, which happen to offset exactly the previous favorable variance on nonlabor due to the travel variance. In the fourth month, the purchase order is paid, and the net variance for the task becomes zero.

Actual work may differ from plan.

Figures 19-2 through 19-7 illustrate that variances occur because of the payment timing and that actual performance differs from plan. They also indicate the necessity of examining the details in project cost reports and commitment reports to understand the reported variances and their significance.

Each computer-based system differs in detail, so the project manager should understand exactly how the reports are prepared (that is, to what errors the reports are prone) as well as the specific meaning of each column of data.

PROJECT COST REPORT

PROJECT____MATERIALS STUDY____ TASK__B-OBTAIN MATERIALS_ DEPARTMENT___CHEMICAL___

	CATEGORY	MONTH 3 - MAR			TOTALS TO DATE		
		PLAN	ACTUAL	VARIANCE	PLAN	ACTUAL	VARIANCE
HOURS	SR. PROF.	2	3	⟨1⟩	46	47	⟨1⟩
	JR. PROF.						
	SR. TECH.						
	JR. TECH.						
DOLLARS	LABOR	100	125	⟨25⟩	1,150	1,175	⟨25⟩
	OVERHEAD	100	125	⟨25⟩	1,150	1,175	⟨25⟩
	NONLABOR	0	0	0	4,400	350	4,050
	PRIME COSTS	200	250	⟨50⟩	6,700	2,700	4,000
	G & A	30	38	⟨8⟩	1,005	405	600
	TOTAL COSTS	230	288	⟨58⟩	7,705	3,105	4,600

FIGURE 19-6. *Cost report for third month for task B work by the chemical group.*

PROJECT COST REPORT

PROJECT_____MATERIAL STUDY_____ TASK __B-OBTAIN MATERIALS__ DEPARTMENT____CHEMICAL____

CATEGORY		MONTH __4 - APR__			TOTALS TO DATE		
		PLAN	ACTUAL	VARIANCE	PLAN	ACTUAL	VARIANCE
HOURS	SR. PROF.	0	0	0	46	47	⟨1⟩
	JR. PROF.						
	SR. TECH.						
	JR. TECH.						
DOLLARS	LABOR	0	0	0	1,150	1,175	⟨25⟩
	OVERHEAD	0	0	0	1,150	1,175	⟨25⟩
	NONLABOR	0	4,000	⟨4,000⟩	4,400	4,350	50
	PRIME COSTS	0	4,000	⟨4,000⟩	6,700	6,700	0
	G & A	0	600	⟨600⟩	1,005	1,005	0
	TOTAL COSTS	0	4,600	⟨4,600⟩	7,705	7,705	0

FIGURE 19-7. *Cost report for fourth month for task B work by the chemical group.*

Variances Due to Overhead Rate Changes

Figure 19-8 is the project report for task E in the technical support section, as reported at the end of the fourth month. In this report, there are favorable variances in labor but an unfavorable variance in overhead. How can this be? If labor is favorable, why should overhead be unfavorable?

Figure 19-9 shows the cause of this. A planned overhead rate (namely, 100 percent) was based on a planned workload for the technical support section (or perhaps the entire research and engineering division). But the actual overhead rate at the end of the fourth month is higher (namely, 130 percent) because the workload base for the entire section has been reduced from plan. The overhead expenses have been somewhat reduced, but not in the same proportion as the direct labor because overhead is partially composed of fixed expenses that cannot be reduced. Thus, the actual overhead rate turns out in this case to be 130 percent rather than the planned 100 percent. (Such a change in overhead is extreme; I use it simply to dramatize the possible effect of overhead rate changes). The variances could be summarized as follows:

Factors outside the manager's control may cause costs to vary from plan.

A senior technician was planned full-time but was not released from his previous project at the start of month 4. If labor is not added, the project will be late.

Overhead is now 130 percent, not 100 percent, as planned. This will cause a cost overrun unless compensating savings can be found.

PROJECT COST REPORT

PROJECT____MATERIALS STUDY____ TASK__E-EXPERIMENT____ DEPARTMENT_TECHNICAL SUPPORT

CATEGORY		MONTH __4 - APR__			TOTALS TO DATE		
		PLAN	ACTUAL	VARIANCE	PLAN	ACTUAL	VARIANCE
HOURS	SR. PROF.						
	JR. PROF.						
	SR. TECH.	200	120	80	200	120	80
	JR. TECH.	400	400	0	400	400	0
DOLLARS	LABOR	7,000	5,800	1,200	7,000	5,800	1,200
	OVERHEAD	7,000	7,540	⟨540⟩	7,000	7,540	⟨540⟩
	NONLABOR	0	0	0˙	0	0	0
	PRIME COSTS	14,000	13,340	660	14,000	13,340	660
	G & A	2,100	2,001	99	2,100	2,001	99
	TOTAL COSTS	16,100	15,341	759	16,100	15,341	759

FIGURE 19-8. *Cost report for fourth month for task E work by the technical support section.*

Planned

Project Example $100,000 Labor Cost	Project Otherone $300,000 Labor Cost
Overhead ($200,000 Fixed + $200,000 Variable) $400,000 = 100%	

Revised

Project Example $100,000 Labor Cost	Project Otherone $150,000 Labor Cost
Overhead ($200,000 Fixed + $125,000 Variable) $325,000 = 130%	

FIGURE 19-9. *Cause of unfavorable overhead variance.*

PROJECT COST REPORT

PROJECT_____**MATERIALS STUDY**_____ TASK_**E-EXPERIMENT**_____ DEPARTMENT_**TECHNICAL SUPPORT**

CATEGORY		MONTH __5 - MAY__			TOTALS TO DATE		
		PLAN	ACTUAL	VARIANCE	PLAN	ACTUAL	VARIANCE
HOURS	SR. PROF.						
	JR. PROF.						
	SR. TECH.	40	20	20	240	140	100
	JR. TECH.	80	120	⟨40⟩	480	520	⟨40⟩
DOLLARS	LABOR	1,400	1,500	⟨100⟩	8,400	7,300	1,100
	OVERHEAD	1,400	1,950	⟨550⟩	8,400	9,490	⟨1,090⟩
	NONLABOR	0	0	0	0	0	0
	PRIME COSTS	2,800	3,450	⟨650⟩	16,800	16,790	10
	G & A	420	518	⟨98⟩	2,520	2,519	1
	TOTAL COSTS	3,220	3,968	⟨748⟩	19,320	19,309	11

FIGURE 19-10. Cost report for fifth month for task E work by the technical support section.

Figure 19-10 is the project cost report for this task in the technical support department at the end of the fifth month. The senior technician category continues to have a favorable variance, which is partially offset by an unfavorable variance in the junior technician category and the continuing unfavorable variance in the overhead. The overhead variance is again attributable solely to the overhead rate now being 130 percent rather than 100 percent, as per plan. Examination of these variables might lead to the following kind of information:

The senior technician is sick, and this category of labor is below plan.

A junior technician, previously unplanned, has been added and has been able to complete the technician work on schedule.

Overhead is still 130 percent, but final dollars are okay.

The junior technician was able to accomplish in fewer hours the work previously planned for a senior technician (which can indeed happen). This net effective saving in labor hours was sufficient to compensate for the increased overhead dollars. The end result in this case is a small, favorable variance.

COST/SCHEDULE CONTROL SYSTEM CRITERIA

The Cost/Schedule Control System Criteria were devised by the U.S. Department of Defense to predict cost overruns early in major military procurement projects

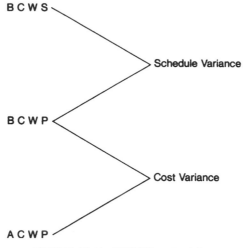

FIGURE 19-11. *C/SCSC methodology.*

and thus reduce risk. NASA has also adopted this approach for some projects. Fundamentally, C/SCSC is a form of earned value, which has been widely emulated on other types of projects. On many projects, the actual cost being reported conformed closely to the initial planned cost, at least during the initial months, but this concealed the fact that actual work accomplishments were lagging behind plan. Consequently, the total project cost for the work performed exceeded planned cost for the work accomplished. Thus, when projects were far advanced and major commitments had been made to them, it became apparent that there were going to be major cost overrun problems at the end for which no contingency reserve existed.

As with many things done by the military, this is a very formal, rigid system. It provides a standard terminology and approach, which clearly are helpful, but it saddles both the military and its contractors with a highly detailed system. C/SCSC must be initiated in the planning phase, but it is really more useful as a control tool than as a planning device.

C/SCSC is used on major military projects.

To work with this system, you must deal with three quantities, shown in Figure 19-11. The budgeted cost of work scheduled (BCWS) is compared with the budgeted cost of

|- - - - - - -| = One-month duration

|- - - - - - -|- - - - - - -|- - - - - - -| = Total task

3   3   3

FIGURE 19-12. *A three-month task with planned costs of $3,000 per month.*

|- - - - - - -|- - - - - - -|- - - - - - -| = Option A

   1   3   5

|- - - - - - -|- - - - - - -|- - - - - - -| = Option B

   3   3   3

|- - - - - - -|- - - - - - -|- - - - - - -| = Option C

   5   3   1

FIGURE 19-13. *Three options for describing your plan for use with an earned value measurement system (for instance, C/SCSC).*

work performed (BCWP) in a given period. The difference between these is a measure of schedule variance. In addition, the actual cost of work performed (ACWP) is compared to the budgeted cost of the work performed, and the difference, if any, is a cost variance.

To get a sense of how this system works, imagine that you have a three-month task and that you expect it to require a uniform rate of effort with costs equal to $3,000 per month. Also imagine that the task could be divided into three one-month subtasks, as shown in Figure 19-12. Figure 19-13 portrays three options you might use to describe your plan for use with an earned value system (for instance, C/SCSC). Table 19-1 summarizes the three quantities and the two variances for each option at the end of each month. In option A, a so-called back loaded project plan, you have unfavorable variances until the end. In option C, a so-called front

TABLE 19-1 Analysis of Earned Value Exercise Options

Option	BCWS	BCWP	ACWP	Schedule	Budget
A					
End of Month 1	1	1	3	0	−2
End of Month 2	4	4	6	0	−2
End of Month 3	9	9	9	0	0
B					
End of Month 1	3	3	3	0	0
End of Month 2	6	6	6	0	0
End of Month 3	9	9	9	0	0
C					
End of Month 1	5	5	3	0	+2
End of Month 2	8	8	6	0	+2
End of Month 3	9	9	9	0	0

loaded project plan, the situation is favorable until the end. This illustrates that the kinds of estimates used in constructing the plan can conceal schedule and cost variances. In short, the system is still subject to manipulation by a clever project manager. The details of the C/SCSC system are far too intricate to explain here and are of only limited interest. If you must use it, seek specialized training.

COMPUTER SOFTWARE

The possibility of comparing actual versus planned costs is a frequently advertised feature of computer software. As I said earlier, it is relatively easy to insert expected costs (labor and nonlabor) for each task when it is entered into a computer project management software package. It is also possible during project performance to enter actual costs information. However, most large corporations collect these actual costs data in the corporate mainframe computer (or a dedicated project management mini-computer). That is, each cost center (for instance, different departments) working on the project provides data that are transmitted from their location (often remote from the project manager). Although it is possible to get a report from the corporate computer and enter these data in your microcomputer, it is simpler to have an electronic data link, as illustrated in Figure 19-14. In fact, several project management computer software packages will trade data back and forth with certain project management systems.

Many project management computer software packages (including Microsoft Project 4.1) do not have provision for linkages to a corporate cost collection system. In such a case, a project manager must decide whether to compare actual versus planned cost using the project management software or a simple spreadsheet. The choice will

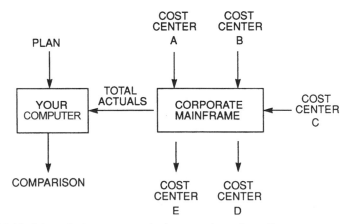

FIGURE 19-14. *It is easier to compare actual versus planned cost if your corporate project cost accounting system can provide actual cost information via an electronic data link to your project management computer software.*

This material originally appeared in a videotape presentation produced by the Association for Media-based Continuing Education for Engineers, Atlanta, Georgia.

be a personal one. (I favor the use of a spreadsheet because there are more clerical personnel who are familiar with these who could handle much of the routine data entry.)

Other project management computer software packages (including Microsoft Project '98) can be linked to the corporate computer that is being used to capture project cost charges. The technical details of how to make this link work depend on the software being used. Where such a link can be constructed, the project manager can have nearly real-time cost data and need not spend time tabulating data from incompatible systems.

TYPICAL PROBLEMS

Cost reports are late and sometimes incorrect.	In addition to management's tendency to presume that cost status is the only measure of a project's progress, a problem previously mentioned, there are two other common problems. First, cost reports are never available immediately after the end of the reporting period (typically the fiscal or calendar month). They are usually issued about two or three weeks after the close of the report period. Thus, there is always a time lag with which the project manager must contend. This problem cannot be solved but must be accepted because of other accounting priorities. An alert project manager can recognize the situation and make full use of other available data to stay more current, including weekly labor distribution reports (indicating who did or did not charge time to the project), purchase orders, travel vouchers, and drawing releases.

The second problem is that cost reports often have errors, such as charges that should have been allocated to other projects or overhead accounts. The project manager or an administrative assistant must study the reports carefully and not merely accept them as gospel. |

MATERIALS STUDY PROJECT

All the examples in this chapter have been based on the materials study project. The ability of computers to sort stored data in a wide variety of ways makes it easy to provide status reports to managers. A senior manager, for instance, Ed Ross, could review other summary information of specific interest to him. Similarly, support team managers could get reports on their subordinates' work.

HIGHLIGHTS

- Computer-generated cost reports show variances from plan, which usually require corrective action.
- Reports can show variances due to timing, actual work deviating from plan, or overhead rate changes.
- Three problems with cost reports are that management tends to assume they are the only measure of progress, they never appear immediately after the reporting period, and they may contain errors.

20

Handling Project Changes

Change is a reality of project management. This chapter first reviews why project plans are altered and then discusses techniques for making changes.

REASONS FOR CHANGES

Change is a constant on projects.

Deviations from the project plan occur because Murphy was an optimist. Or there is Rosenau's law of revolting developments: There will be at least one. A large body of data supports this assertion. Table 20-1 summarizes overrun data from a variety of sources. These studies all had a nominal correction for inflation, which is why I selected them. Although the averages show overruns, some projects were completed early and underrun; but, of course, some were much worse than the reported averages. These data convincingly demonstrate that presumably conscientious project managers have been confronted with changes in many situations. This has led to missing both the schedule and cost dimensions of the Triple Constraint, with the overruns typically being a factor of approximately two above plan. In other situations, the performance goal may prove to be unattainable.

There may be more recent overrun studies of which I am not aware. A more widespread knowledge of project management principles and better software should produce better project outcomes. But projects continue to both overrun and underrun. For example, the Auger deep drilling platform in the Gulf of Mexico was plagued by cost overruns and delays before final completion in 1994. Conversely, the subsequent Mars deep drilling platform, estimated to cost $1 billion, was completed in 1996 with an underrun of 20 percent. The essential point is that projects encounter changes and are likely to be completed differently than planned.

Estimates for smaller projects are more accurate.

Although not illustrated in Table 20-1, Mansfield's data also show that the situation is somewhat worse than average for large projects and better than average for small projects. This seems to make sense; intuition tells us that the more ambitious undertaking is less likely to be estimated accu-

TABLE 20-1   Time and Cost Overrun Data, Expressed As Multiple (X) of Plan

Project Type	Time	Cost	Source
50 new products (new chemical entities, compounded products, or alternative dosage forms) in ethical drug firm	1.78X	1.61X	E. Mansfield et al., *Research and Innovation in the Modern Corporation*, W. W. Norton, New York, 1971, p. 89
69 new products in proprietary drug laboratory	2.95X	2.11X	E. Mansfield et al., op. cit., pp. 102 & 104
20 management information systems projects	2.10X	1.95X	R. F. Powers & G. W. Dickson, "MisProject Management? Myth, Opinions, and Reality," *California Management Review*, XV, no. 3, 147–156 (Spring 1973)
34 Department of Defense systems from "planning estimate"	—	2.11X	G. R. McNichols, D.O.D. Report (November 1974), as quoted in R. A. Brown, "Probabilistic Models of Project
from "development estimate"	—	1.41X	Management with Design Implications," *IEEE Trans. Engr. Mgmt.*, vol. EM-25, no. 2, 43–49 (May 1978)
10 major construction projects completed 1956–1977	—	3.93X	W. J. Mead et al. (1977), as quoted in E. W. Merrow et al., "A Review of Cost Estimation in New Technologies," Rand Corporation Report R-2481-DOE (July 1979), p. 38
10 energy process plants	—	2.53X	E. W. Merrow et al., op. cit., p. 87

rately. This is another reason to break a large project down into many small tasks. It will be easier to estimate a small task accurately. You also have nature on your side in the form of the central limit theorem. This theorem predicts that the percentage error will be smaller if a larger number of estimates, each with small random errors above or below the estimate, are made.

Real Changes in the Planned Project

One of the causes of both time and cost overruns is that deviations from the plan occur in the job. They may be externally imposed, by the customer, for instance. They might include a request for increased accuracy in a sensor, a request for extra flexibility in a computer program, a request to add traffic islands on a bridge, or perhaps a request for an additional flight experiment on a space payload. The customer may be late getting you essential information, as one project manager complained:

> In order to complete a major project, I need important and timely data from my customer. Even when the customers do not provide the data, they still demand that I complete the project in a . . . timely manner.

In addition, schedule changes can be imposed, and these frequently have attendant cost implications. If the customer's plant is shut down by a strike or for some other reason the customer does not wish to receive the project output on a stipulated date, there is typically a cost consequence. Conversely, the sponsor may impose a change in project funding. Although a delay in project funding to the project organization may not appear to change the total available funding, it almost always is accompanied by a schedule rearrangement, which normally leads to undesirable cost consequences.

Changes can be imposed indirectly.

Environmental, health, and safety regulations that change during the course of a project may cause other changes in scope. Inflation may exceed plan, causing the cost problem, particularly on projects originally planned to take several years. It is almost axiomatic that there will be changes in resource availability, either people or facilities. Perhaps the person assigned to a key task is not performing as expected, which is a common problem illustrated by another project manager's comment:

> The assigned person charged over two weeks time to my project and never even installed the compiler on her machine. I was constantly asking her about her progress, but she always managed to make it sound like things were moving along.

These do not constitute changes in project scope, but they do constitute changes from plan that will have an impact, usually unfavorable, on schedule and cost.

Estimation Inaccuracy

Several factors affect estimation accuracy. Of these, an imperfect definition of project scope is the most common. Either a sponsor or a performer may be the cause of the error, but it usually is attributable to both.

There may also be poor estimates of either time or cost. The rush to prepare a proposal and submit it in accordance with the bidding requirements may preclude there being sufficient time to do a good job of estimating. There is so much inherent uncertainty in some tasks of some projects that a poor estimate is almost a foregone conclusion.

Many jobs are proposed with deliberate underestimates of the amount of time or money it will take to perform them. This is the so-called "buy-in" situation in which a bidder attempts to win a job by making a low bid. This does not require an illegal misrepresentation, although that may be the case. It may result from a deliberate attempt to make optimistic assumptions about all the uncertainties in the proposed project as well as to omit all contingency from the estimates. In a sense, the bidder is making an estimate of time and cost that could occur perhaps 1 or 0.1 percent of the time rather than attempting to make an estimate near the mid range of possibilities.

Uncertainty and "buy-ins" can cause poor estimates.

Buy-in bids are much more prevalent where the contemplated contract will be a cost reimbursable form and the bidding contractor will not bear the financial burden of having made a low bid. They can also occur in a fixed price contract situation where the bidding contractor is confident that the customer will request changes in scope. Such changes will provide a "get well" opportunity: Increases in both time and cost for the main project can be added onto or concealed in renegotiations necessitated by changes of scope requested by the customer.

ADOPTING CHANGES

At this point, it should be clear that projects normally require changes of plan, although the specific reason or reasons cannot be forecast. The original plan should have included contingency for each dimension of the Triple Constraint, as discussed in Chapter 11. The next chapter reviews several methods to solve the problems caused by the occurrence of revolting developments, stressing the importance of developing and considering alternatives. Once the best alternative has been identified, the change must be adopted.

Figure 20-1 is another variation of a task authorization form, which allows for changes. As I said in Chapter 6, each organization has its own detailed version, but the essential elements are a description of the Triple Constraint, which defines the task being revised, and a place for the person authorizing the revised work and the person accepting responsibility for the revised work to sign. This form thus constitutes a "contract," defining in writing the agreement reached to authorize a revised task. In the case of a subcontractor, a revised or changed subcontract document authorizes the revised task.

Changes, just like originally intended work, must be defined, planned, managed, and monitored before they can be completed. Thus, some, if not all, of the originally issued task authorizations must be changed when a change has occurred and a decision has been made as to how to alter the plan to carry out the remainder of the project. This may seem to be a lot of work, but it is far less onerous to take the time to make sure each agreement with people working on the project has been changed than to discover later that some people have been working according to their prior understanding of the project plan.

Be certain that all team members know of plan changes.

There is a natural reluctance to make a formal plan change. Such a change not only requires work to issue plan revisions, but it forces us to admit we were wrong (in the original plan), and often brings this to the attention of higher management. Conversely, such information, if given to higher management, may result in your project receiving more help, better access to resources, or higher priority.

Never hesitate to publicize plan changes if these are required.

Figure 20-1 shows a one-page form, but an actual task authorization, in common with a subcontract, might be many pages long. The form might be part of a carbon

TASK AUTHORIZATION		PAGE OF
TITLE		

PROJECT NO.	TASK NO.	REVISION NO.	DATE ISSUED

STATEMENT OF WORK:

APPLICABLE DOCUMENTS:

SCHEDULE

START DATE: COMPLETION DATE:

COST:

	ORIGINATED BY:	DATE:	ACCEPTED BY:	DATE:
ORIGINAL	APPROVED BY:	DATE:	APPROVED BY:	DATE:
	APPROVED BY:	DATE:	APPROVED BY:	DATE:
	ORIGINATED BY:	DATE:	ACCEPTED BY:	DATE:
REVISION 1	APPROVED BY:	DATE:	APPROVED BY:	DATE:
	APPROVED BY:	DATE:	APPROVED BY:	DATE:
	ORIGINATED BY:	DATE:	ACCEPTED BY:	DATE:
REVISION 2	APPROVED BY:	DATE:	APPROVED BY:	DATE:
	APPROVED BY:	DATE:	APPROVED BY:	DATE:

FIGURE 20-1. *A task authorization form to authorize take revisions.*

set, providing copies for the initiator, the task manager, and the project cost accounting section.

Hence, the task authorization documents initially used to authorize work are also a major change control document. A large project may generate many of these, and the amount of time it takes to issue them may be so great as to advise that there first be telephonic or other speedy notice of forthcoming changes.

COMPUTER SOFTWARE

If the changes that have occurred on your project do not alter the logic of the plan, merely the duration or cost of each activity, it is easy to use computer software to determine and visualize the impact. Conversely, when the changes require a substantial revision of the plan's logic, it may be more trouble to revise the plan in the computer than to start afresh and merely plan the remaining work. It can be very time consuming to attempt to disconnect task linkages that have previously been entered and replace these with new tasks with a different logical relationship between them.

TYPICAL PROBLEMS

If you change project (or task) managers during a project, you will almost always be confronted with changes, typically unfavorable. The new person may be unfamiliar with the ultimate user environment, although that is not always the case. Similarly, the new manager may not be familiar with team members' capabilities. The overriding issue, however, is that the new project or task manager did not plan the work for which he or she is now responsible—this is a violation of the Golden Rule. If you must change the project or task manager during the course of the activity, allow that person some time to review the prior plan and try to determine how to carry out the remainder of the work.

There is always a reluctance to tell the customer and your boss that a revolting development (such as the discovery of unexpected noise in an amplifier) has occurred and many reasons to justify delay. But you should deliver the bad news carefully, thoughtfully, and promptly, before someone else does it.

A second problem is that task authorizations are often verbal rather than written. Because they promote misinterpretation, verbal authorizations should be avoided. But they are employed in the real world of project management. When you must use them, be sure you are clear, ask for feedback, and then try for written confirmation.

The third problem with changes is their impact on resource allocation. There is nothing to do but face up to the reality that resources must be rescheduled, as inconvenient as this may be.

MATERIALS STUDY PROJECT

In the case of the materials study project, NERESCO might request changes if they wanted extra materials to be evaluated, a duplicate experimental setup delivered for their own use, or a formal briefing at the end of the job.

(continued)

Further changes might arise because a revolting development has occurred. Perhaps the experimental effort (task E) on the materials study project has run into difficulty because one of the materials was found to be impure. If SUPROMAC is anxious to obtain follow-on work from NERESCO, there will be a natural reluctance to inform the customer. Viable alternatives that avoid that necessity would thus be desired. In this instance, properly purified materials can be obtained quickly, and means to recover lost time must then be devised. Two possibilities are using overtime and adding a second shift. If either of these actions can be taken, and the original schedule can be maintained, SUPROMAC need not inform NERESCO.

HIGHLIGHTS

- Changes will occur on every project.
- Changes may result from customer requests; extenuating circumstances, such as a strike; altered environmental, health, or safety regulations; inflation; or resource unavailability.
- Uncertainty and buy-ins can result in inaccurate estimations.
- Authorization documents can be used to communicate planning and change control.
- Three problems changes can cause are managers may be reluctant to inform the customer and higher management of them, verbal authorizations often cause misunderstanding, and resources must be reallocated.

21

Solving the Inevitable Problems

Failure or an unsatisfactory project outcome can result from bad solutions to problems, which are often decisions made hastily or without adequate information. This chapter is concerned with how to solve problems as your project encounters changes during execution. First I discuss some general approaches to coping with problems. Then I describe decision trees, a powerful analytic technique for problem analysis. Following that, I review use of matrix arrays and then discuss the kinds of meetings in which problem-solving approaches are most likely to be effective.

THE GENERAL APPROACH

Good solutions require a seven-step approach.

In general, the options available are either deductive or inductive logic. In the former, the solution is derived by reasoning from known scientific principles, using analytical techniques, and the conclusions reached are necessary and certain if the premises are correct. In practice, the project manager is rarely confronted with problems for which this approach is appropriate and must rely on inductive techniques, for which the scientific method is the typical prototype. Inductive methods reach conclusions that are probable. This straightforward approach entails seven steps, described in the following sections.

State the Real Problem

The key to problem solving is understanding the real problem rather than the apparent symptoms. Smoke may be emerging from the hardware you built or the computer may refuse to obey a subroutine command, but the actual problem may be an overheated component or an improper line of code in the computer program. You will have to decide how and perhaps why these particular problems occurred.

Gather the Relevant Facts

A fact-gathering phase is usually necessary to clarify the problems. People trained in engineering and science tend to want to engage in this step ad infinitum. This has been called "analysis paralysis." Although it may take a good deal of time to locate information sources, there is also a law of diminishing returns. Because you will never have a 100 percent certainty of obtaining all the information, you must learn to exercise judgment as to when to truncate a search for additional information. At that point, you begin to converge on a solution using the information already gathered.

Propose a Solution

Once a plausible or possible solution has been identified, the winners and losers rapidly separate in their approach to problem solving. The losers inevitably adopt the first solution that comes to mind, possibly leaping out of the frying pan and into the fire. Admittedly, the pressures to come up with a solution quickly are great. No one likes to walk into his or her boss's or customer's office and say, "We have a problem." Such a crisis generates psychological pressure in the project manager to come up with a solution quickly so he or she can say, "We have a problem, but don't worry about it because we have a pretty good solution in mind." But it is best to take a different approach. As stated by Emile Auguste Chartier, "There is nothing more dangerous than an idea when it is the only one we have."

Develop Several Alternative Solutions

The winning approach to problem solving is to develop several alternative solutions. Thus, when the problem has arisen and must be reported, the successful project manager will say, "We have a problem. We may have a possible solution, but I am going to take three or four days to consider other alternatives. Then I will report to you on the options

Developing alternatives is the key to problem solving.

and our recommended course of corrective action." Although such an approach to reporting the bad news may make you initially uncomfortable, it is invariably associated with reaching better solutions.

Adopt the Best Alternative

After deciding what is the best alternative, you must adopt a course of action.

Tell Everyone

As an effective project manager, you have earlier made certain that everybody involved in your project knew the original project plan. Now, because you have changed one or more dimensions of the Triple Constraint, you must tell everyone what the new plan is. If you fail to do this, there will be some people working in accordance with obsolete direction, producing something useless and out of date.

Audit the Outcome

As you implement the best alternative solution, watch how it it is working out. Auditing will improve your ability to solve problems by showing you how your solutions actually work out. And as you learn more about the problem you are solving and the approach you have adopted, a better alternative may become clear, which may necessitate a further change in the plan.

DECISION TREES

Choosing the best alternative often requires estimating the possible outcomes and their probabilities. An organized way to cope with the situation is to use a decision tree. This technique is both simple and powerful. If you use it often enough, it will improve your average performance in adopting alternatives. Consider the decision whether to go to a movie or walk on the beach. This decision and its possible outcomes are shown in Figure 21-1.

> **Decision trees require you to quantify the outcome.**

Each possibility has chance future events, which have to do with the quality of the movie or the weather. There are also different outcomes, which are illustrated. Outcomes such as those shown, which have more than one dimension to them (happiness and money, in the figure) must be reduced to a single numeric value. This can be done by utility or preference techniques.

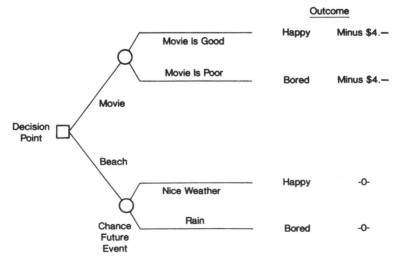

FIGURE 21-1. *A decision tree.*

Figure 21-2 is a general representation of the kinds of decision trees with which you must work. A decision tree always starts with a decision for which there are two or more possible choices. Each choice may be followed by chance future events or subordinate choices (decisions) in any order and with two or more branches following each node. There is a single numeric value outcome, which is typically the present value of the cash flows along the various branches to that outcome.

Decision trees are used routinely in many situations, such as the decision whether to drill an oil well at a given location or whether to locate a new warehouse in a new geographic region. In both of these typical situations, the correctness of the decision depends on the probabilities of things happening in the future (oil being found or business growth), and decision trees are designed to maximize the likelihood of choosing the correct course of action.

Consider the following situation, which often confronts a project manager. You have just received an unsolicited request for a proposal that will yield a $300,000 before-tax profit if you can win the job. Checking with the marketing manager, you learn that your company and two others were both solicited suddenly and you all have an equal opportunity of winning the job. Thus, you appear to have one chance in three of winning $300,000 if you write the proposal.

After discussing this, you and the marketing manager realize you have the opportunity to construct a working model for $45,000 and doing so increases the odds of your winning to fifty-fifty. The decision with which you are confronted is whether to build a working model (at a cost of

There are two ways to draw decision trees.

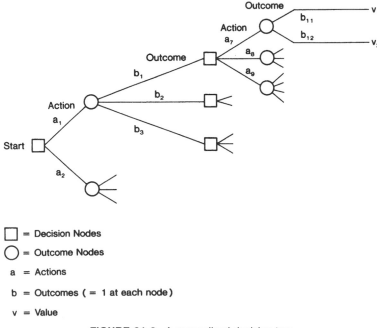

= Decision Nodes

○ = Outcome Nodes

a = Actions

b = Outcomes (= 1 at each node)

v = Value

FIGURE 21-2. *A generalized decision tree.*

$45,000) to increase the odds of winning a $300,000 before-tax profit from one in three to one in two. Figure 21-3 shows the decision tree and the problem analysis for this solution. Figure 21-4 is an alternative representation. In both cases, the branch in which the model is built has a higher value than the branch in which the model is not built. Therefore, your decision would be to build the model because it has a higher expected value. Note that this does not guarantee you will win the job. Rather, it gives you a higher expected dollar value. If you use decision trees in enough cases, you will do better over the long run. But you might very well develop the model and still lose the job, thus ending up losing $45,000. (Note that, for simplicity, I have ignored the cost of writing the proposal in both situations.)

Imagine that you have constructed the preceding decision tree (Figures 21-3 and 21-4), but you and the marketing manager continue to discuss the situation prior to building the model. You realize that, whether or not you build the model, if you lose the proposal, you will have an opportunity to write a subsequent proposal to the winning contractor. Such a subsequent proposal will cost you an additional $10,000, but if you write it after losing, you will have an 80 percent chance to win a subcontract (worth $100,000 in before-tax profit) from the winning contractor. In this situation, you would decide not to build the model, as illustrated in

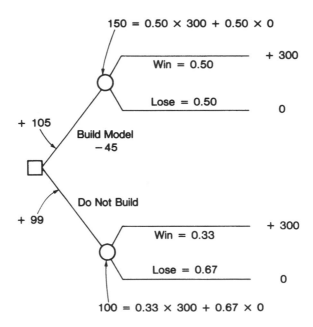

Methodology:
  • Construct diagram from left to right.
  • Insert $ values from right to left.

FIGURE 21-3. *Decision tree for illustrative example (thousands of dollars omitted).*

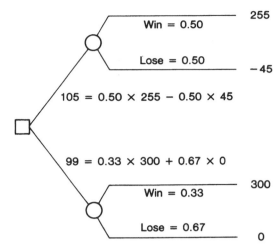

FIGURE 21-4. *Alternative representation of Figure 21-3 decision tree.*

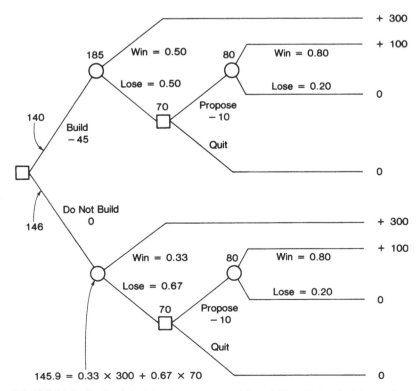

FIGURE 21-5. *Revised decision tree to account for additional later decision options.*

Figure 21-5. (Another way to look at this revised situation is to realize that losing the original job is less serious, having an expected value of $70,000 instead of zero.)

Sometimes you are confronted with decision choices, for example, how to remove a resource constraint on two projects (as indicated in Figure 10-4), for which a qualitative decision tree (Figure 21-6) is useful. In this, you may be unable to quantitatively estimate outcome values, odds of occurrence, or costs of actions. However, you may be able to clarify the alternatives to the point where a better choice is possible.

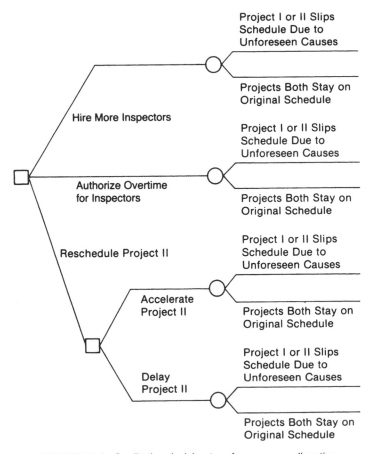

FIGURE 21-6. *Qualitative decision tree for resource allocation.*

MATRIX ARRAY

Quantitative

In situations where decision trees are not practical or are unwieldy for the analysis and comparison of alternatives, the matrix array may be a satisfactory aid. Figure 21-7 illustrates a quantitative weighting (or scoring) array. The key considerations in the particular problem are listed on

A matrix must list all significant criteria.

the left margin. In a computer programming project, the performance criteria might include processing speed, size of memory required, or similar considerations. For an airborne piece of equipment, performance issues might be weight, size, and reliability. In addition to performance targets criteria, the listing should include the schedule and cost implications of adopting that particular solution.

Next, weighting factors (the sum of which is equal to one) are attached to each criterion (or target). Then a percentage, indicating the degree to which the solution satisfies the particular criterion target, is entered for each solution. Finally, each percentage for each solution in the body of the matrix is multiplied by the corresponding weighting factors, and the result is entered at the bottom of the solution column (for example, for solution approach 2, $0.20 \times 90\% + 0.15 \times 70\% + 0.15 \times 50\% + 0.20 \times 70\% + 0.30 \times 80\% = 74\%$). There are four problems with this approach:

Criteria	Weighting Factor	Solution Approach 1	Solution Approach 2	Solution Approach 3
Performance Target A	0.20	60%	90%	80%
Performance Target B	0.15	90%	70%	70%
Performance Target C	0.15	90%	50%	90%
Schedule Target	0.20	70%	70%	90%
Cost Target	0.30	90%	80%	70%
Weighted Percentage Value		80%	74%	79%

FIGURE 21-7. *Quantitative decision matrix.*

1. Deciding on weighting factors may be difficult, and a strong-minded person may easily argue for or manipulate these to favor his or her preferred option.
2. Choosing percentages may be difficult, especially for the schedule and cost targets, where the solution approach may exceed plan.
3. The highest ranking or weighted percentage value may still be inadequate.
4. It does not consider people, in particular, the possibility that one solution approach is championed by an ambitious person who will work nights and weekends to accomplish it.

Computer software is available to mechanize this kind of approach, although a spreadsheet seems to be perfectly adequate. A new form of software—the so-called "fuzzy logic" decision matrix—is also useful. The analytic hierarchy process is one commercial embodiment. In distinction to the purely quantitative matrix, this approach replaces the need for choosing weighting factors with the need to choose relative importance ratings. The potential advantage is that this is less prone to advocacy that can favor a forceful person's preferred choice.

Figure 21-8 is an example of a fuzzy logic decision using data from the matrix of Figure 21-7. The weighting factors from Figure 21-7 have been replaced by paired comparisons of criteria using the numerical scale of Table 21-1. As an example, two lines below "RATINGS OF THE CRITERIA" in Figure 21-8, you see a "1" and a "4." This means that criterion 1 of the pair on the previous line (Performance Target A) is chosen as having somewhere between a "strong" and a "weak" importance over criterion 2 (Performance Target B). In the last pair of lines in that section of Figure 21-8, the Cost Target criterion has been chosen as having a strong importance over the Schedule Target. As shown in Figure 21-8, these data indicate that Solution Approach 1 is preferred. Because the computational algorithm is obscure, it is harder for assertive persons to manipulate fuzzy logic to favor the result they desire.

Qualitative

Because of the manipulation to which the quantitative approach is subject, it is often better to use a slightly more qualitative matrix. Such a matrix can explicitly consider people issues, especially where there are advocates for or opponents to a particular solution approach. Consider the following problem a project manager called to my attention:

> An employee volunteered to perform fieldwork at a remote off-site location on my project and accepted relocation to that site for a one-year tour of duty. After two months, the field site manager reports that the employee is not getting along with the rest of the crew, and the employee calls and says that he wants to return. At this point, there are no other qualified employees who are willing to relocate.

In this kind of situation, there are really no tractable, quantitative approaches. In this kind of situation, a qualitative approach is all that can be used.

RATINGS OF THE ALTERNATIVES
```
.6  .9  .8
.9  .7  .7
.9  .5  .9
.7  .7  .9
.9  .8  .7
```

RATINGS OF THE CRITERIA

| (1) Performance Target A | — | (2) Performance Target B |
| 1 | 4 | |

| (1) Performance Target A | — | (2) Performance Target C |
| 1 | 4 | |

| (1) Performance Target B | — | (2) Performance Target C |
| 1 | 1 | |

| (1) Performance Target A | — | (2) Schedule Target |
| 1 | 1 | |

| (1) Performance Target B | — | (2) Schedule Target |
| 2 | 4 | |

| (1) Performance Target C | — | (2) Schedule Target |
| 2 | 4 | |

| (1) Performance Target A | — | (2) Cost Target |
| 2 | 5 | |

| (1) Performance Target B | — | (2) Cost Target |
| 2 | 7 | |

| (1) Performance Target C | — | (2) Cost Target |
| 2 | 7 | |

| (1) Schedule Target | — | (2) Cost Target |
| 2 | 5 | |

FDM PROGRAM OUTPUT
****

EIGENVALUE = 5.084712

EIGENVECTOR . . .
  .1639196   5.148857E-02   5.148857E-02   .1639196   .5691838

ALPHA-VECTOR . . .
  .8195978   .2574428   .2574428   .8195978   2.845919

CONSISTENCY OF THE PAIRED-COMPARISON MATRIX = .1029029
(SHOULD BE LESS THAN 1.0)

WEIGHTED FUZZY SETS . . .
```
.6579203    .9172701    .8328615
.9732402    .9122662    .9122662
.9732402    .8365695    .9732402
.7465221    .7465221    .9172701
.7409311    .5299099    .3623778
```

DECISION VALUES . . .
Solution Approach 1 -   .6579203
Solution Approach 2 -   .5299099
Solution Approach 3 -   .3623778

Solution Approach 1 IS THE BEST CHOICE ACCORDING TO THE DATA YOU HAVE ENTERED.

FIGURE 21-8. *One example of output from a fuzzy logic decision matrix software program.*

TABLE 21-1 Explanation of Numerical Scale Used for Ratings of Criteria in Figure 21-8

Numerical Rating	Meaning
1 - 2	Equal importance
3 - 4	Weak importance of one over the other
5 - 6	Strong importance of one over the other
7 - 8	Demonstrated importance of one over the other
9 -	Absolute importance of one over the other

A qualitative matrix can use quantitative data where these are available (see Figure 21-9). Note three aspects of this approach:

1. Numbers are used whenever possible.
2. People are given explicit consideration.
3. The summary identifies both favorable and unfavorable issues.

Also, this kind of matrix can always be used, even early in the decision analysis, before all relevant numerical data are available.

Issue \ Solution	1	2	3	4	5	6
A						
B						
Schedule						
Cost						
People						
Favorable						
Unfavorable						

FIGURE 21-9. Qualitative matrix.

Figure 21-10 shows a qualitative matrix array for the problem of selecting a three-axis "frictionless" support for an airborne instrument to allow 5 degrees of movement in three directions. The original perceptions of the factors to be considered are typed in the left margin. The first two solution approaches are listed. Both have many drawbacks. In Figure 21-11, a third alternative solution is identified, pointing out the need to consider one more factor (outgassing). Figure 21-12 shows a reasonably adequate solution for which the schedule must be studied further or perhaps accelerated.

To use this kind of matrix, list each solution across the top. You must do this conscientiously. When you write down the first solution and it clearly falls short with regard to one or more of the key criteria, seek additional solutions designed to overcome the shortfall of the ones presently conceived. Usually, after you have written down two, three, or four solutions, each of which has one or more aspects of shortfall, you will identify some hybrid or new variant that comes close to satisfying all the key considerations identified. Even if the matrix array does not

Be honest about solution shortfalls.

Consideration \\ Approach	Conventional Outer-Ring Gimbal	Internal Gimbal				
Weight (Lbs)	50 ?	20 ?				
Size	Huge	Should Be OK				
Reliability	Bearings	Bearings				
Coercion (In-Lbs)	1 ?	1 ?				
Schedule	Quick	Less Quick				
Cost	Low	Higher				
People						
Pro						
Con	Weight, Size, Coercion	Weight, Coercion				

FIGURE 21-10 Qualitative matrix array.

Approach / Consideration	Conventional Outer-Ring Gimbal	Internal Gimbal	Gas Bearing			
Weight (Lbs)	50 ?	20 ?	10-20			
Size	Huge	Should Be OK	OK, But Also Has Pump			
Reliability	Bearings	Bearings	Pump			
Coercion (In-Lbs)	1 ?	1 ?	Nil			
Outgassing	Oil	Oil	Support Gas			
Schedule	Quick	Less Quick	8-12 Months			
Cost	Low	Higher	Still Higher			
People			Jack			
Pro			Coercion			
Con	Weight, Size, Coercion	Weight, Coercion	Weight, Gas, Reliability, Time			

FIGURE 21-11 Qualitative matrix array with a third alternative solution identified.

lead to a hybrid that clearly satisfies all significant considerations, using the array will clarify available trade-offs and options.

PROBLEM-SOLVING MEETING STYLES

There are four meeting styles for use when solving problems:

1. Command—issue orders
2. Negotiative—different groups horse trading to reach agreement
3. Collegiate—peers reaching decision by consensus
4. Advisory—exchange information and make subsequent decision

And there are five approaches to problem solving. If the information to solve the problem is readily available, a solution-centered approach to problem solving may

Approach / Consideration	Conventional Outer-Ring Gimbal	Internal Gimbal	Gas Bearing	2 Horizontal Flexures + Vertical Mercury	3 Equal (35.2°) Flexures	1 Horizontal + 2 45° Flexures
Weight *(LBS)*	50 ?	20 ?	10-20	∼ 10	5-10	5-10
Size	Huge	Should Be OK	OK, But Also Has Pump	Reasonable	Small	Small
Reiiability	Bearings	Bearings	Pump	Mercury Spillage	∞	∞
Coercion *(In-LBS)*	1 ?	1 ?	Nil	∼ 0.01	∼ 0.01	∼0.01
Outgassing	Oil	Oil	Support Gas	Mercury Vapor	Nil	Nil
Schedule	Quick	Less Quick	8-12 Months	3-4 Months	5-6 Months	4-5 Months
Cost	Low	Higher	Still Higher	∼$15K	∼ $25K	∼$20K
People			Jack	Hy	Mike	Mike
Pro			Coercion	Time & Cost, Coercion	Weight, Size, Reliability, Coercion	Weight, Size, Reliability, Coercion
Con	Weight, Size, Coercion	Weight, Coercion	Weight, Gas, Reliability, Time	Mercury	Harder to Make Parts	Time = ?

FIGURE 21-12 *Qualitative matrix array with one reasonable adequate solution.*

be adopted. If the managerial authority is accepted and the manager has the information and the ability to solve the problem, a directive approach is appropriate, and a command meeting, one in which the manager issues orders, is the most appropriate meeting style. This does not mean that a different meeting style cannot be used, but a command meeting will most likely produce an effective solution under these conditions. The negotiative approach to problem solving is bargaining with different objectives but common interests, and, in this case, a negotiative meeting is most likely to be effective. The prescriptive approach to problem solving is one in which a solution is solicited, and the presented answer may be only a tentative trial. For this, a negotiative or collegiate meeting style is most useful.

There are two problem-centered approaches to problem solving, those is which information is still needed. The first of these is the consultative approach, in which trust exists and information sharing is useful to diagnose the problem. The second is the reflective approach, which is useful if the problem is unclear and nonjudgmental restatements are acceptable. In both of these approaches, the advisory meeting is most likely to be productive.

TYPICAL PROBLEMS

In most project management problem-solving situations, it is not possible to find *the* answer, only a most acceptable (or least objectionable) answer. This may be caused by the inherent uncertainty or lack of quantitative data. It is thus a matter of judgment about when to choose among the identified solutions and when to keep looking for more, better solutions. Honest people will differ (as they will in their perception of the problem and their evaluation of solution alternatives), and this must be both expected and tolerated. Use of a qualitative matrix or a purely qualitative decision tree can reduce, if not eliminate, this problem.

MATERIALS STUDY PROJECT

Considering the materials study project problem from the prior chapter, a qualitative matrix (Figure 21-13) suggests overtime is the least objectionable solution.

Solution / Consideration	Overtime	Second Shift
Performance	OK	OK
Schedule	OK	OK, but must train crew
People	Same as original plan	Need leader and two technicians
Cost	Overtime premium	Shift premium
Pro	People available	Less costly
Con	Overtime premium exceeds shift premium	Extra people training time

FIGURE 21-13 *Qualitative matrix for the materials study project problem cited in Chapter 20.*

HIGHLIGHTS

- The seven steps in problem solving are identify the problem, collect the data, devise a solution, search for alternative solutions, adopt the best solution, implement the solution, and audit the outcome.
- Decision trees help you choose the best alternative.
- Matrices, another aid to alternative selection, may be qualitative or quantitative.
- There are appropriate meeting styles for different problem-solving approaches.

Part 5

Completing a Project

22

How to Complete a Project

The fifth and last managerial activity is project completion. To introduce this topic, this chapter first discusses project life cycles. This discussion reviews activities required to complete the project. Then the consequences of project completion are discussed, and it is shown that all personnel do not necessarily have the same stake in ending the project.

PROJECT LIFE CYCLE

Phases

Figure 22-1 shows a project life cycle. The three phases are arbitrary, but they are sufficient to illustrate the point that project activities change during the duration of any project. This has implications for the illustrated personnel head count and means the kinds of personnel used in different phases will have to change. For instance, creative designers, very useful in the early phases, can easily become an obstacle to completion if they are retained during the later phases.

> *That projects have different phases emphasizes that personnel needs will change throughout the project life cycle.*

Termination

There is a variety of ways to stop projects. Resources can be withdrawn, for instance, by reassigning personnel or required facilities. Higher priority projects may gain at the expense of a low priority project, which may be allowed to wither on the vine. These approaches are not as desirable as an orderly and carefully planned termination. Project success, that is, satisfying the Triple Constraint, can be obtained only by

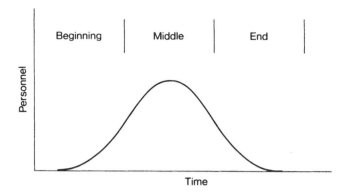

	Beginning	Middle	End
Construction	Site	Erection	Landscaping
Product Development	R&D, Market Research	Engineering and Manufacturing	Product Introduction
Aerospace System	Engineering	Assembly	Customer Test and Sign-off
Computer Software	Definition and Design	Programming	System Test and Acceptance

FIGURE 22-1. *Project life cycle.*

this latter approach. During the last few months of a project, weekly reviews may be required; and during the last few weeks, daily reviews may be required.

Acceptance

Never include unspecified service in a project.

The goal of project management is to obtain customer acceptance of the project result. This means the customer agrees that the performance dimension specification of the Triple Constraint has been met. Unless the acceptance criteria have been clearly defined in documentation agreed to by the customer and the contractor, there will be discord at the end of the project. When agreement is lacking, the customer will typically want more (unless it costs unacceptable amounts of time or money) and the contractor will argue for less. This is why an unstated or undefined amount of service should not be included in a project. The situation between the project sponsor and performers is similar inside a single organization, where one party may have expectations that differ from those of another party.

Therefore, the acceptance phase must start with the initial contractual or otherwise written definition of the work to be undertaken. This is not to say there cannot be changes of scope during the contract to alter or clarify the acceptance criteria. Uncontrolled feature creep ("better" is the enemy of "good enough") is insidious, but a few well-understood and agreed to goal changes, as depicted in Figure 22-2, can be tolerated.

In some projects, it may be impossible at the beginning to agree upon final acceptance criteria. This is typical of high-technology development projects. A related problem is also prevalent in commercial product development projects, where marketing is reluctant to settle on all the performance specifications. When this is the case, the contract or product specification should call for an initial effort of an adequate duration to clarify the entire system or product design and acceptance criteria. At the end of this first phase, a sponsor review is conducted and a contract is negotiated or a firm product specification is agreed to for the final phase of work, including acceptance criteria for the end of the project. This approach requires the sponsor to bear a large risk during the initial phase because the contractor or development team may decide during the initial phase that the final phase will be lengthy and costly. Nevertheless, there is no reasonable alternative to this two-phase approach.

Because there are many possible completion points and delivery conditions for a project, it is necessary to think these through. Completion of the Apollo lunar project would differ depending on whether it was defined as the launch to the moon, the manned landing, the astronaut recovery, or the completion of rock analysis.

Completion requires objective and measurable criteria be attained, which ideally solves the customer's problems.

Objectively measurable criteria for completion (such as relocate this facility and have it ready to operate by next

Goal

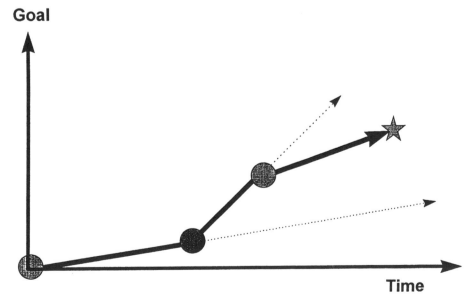

Time

FIGURE 22-2. Scope control requires that necessary goal changes be agreed to and clearly defined.

week or test these two amplifiers for gain using a specified method before 9:30 a.m.) are best. Subjective criteria (such as design an attractive housing for this computer) are risky. The former allow little or no room for ambiguity; in the latter, the customer and contractor could easily have different standards. It is especially important to have objective criteria if there is any likelihood that key customer or sponsor personnel will change between the start and completion of the project. Obviously, this is more likely to happen in a long project than a short project. As one example of the turnover of key project personnel, despite a public law requiring that program managers remain on major Department of Defense acquisitions at least four years, a congressional study in the late 1980s revealed that managers averaged less than two years in these program management roles. Such turnover increases the likelihood of technical problems, schedule delays, and cost overruns, all of which a conscientious project manager wishes to avoid.

It is also important to be clear about what the project output is supposed to accomplish. For instance, three very different results occur when the product performs the specified functions, the product was built according to the specified designs, and the product really solved the user's problem.

Clear acceptance criteria are required.

Project completion clearly depends upon the precise wording of the acceptance criteria. There should be no room for doubt or ambiguity, although in practice this is extremely difficult to accomplish at project inception. Both parties may enter into a contract with goodwill, but the contractor may have assumed and perhaps even included in the contract the jointly agreed priority to use customer facilities to validate product output. Such a priority can easily become an ambiguous issue toward the end of the project. Other difficulties in wording arise when the word "appropriate" is used. Appropriate tests or demonstrations that seem clear and simple at the beginning of a project have a way of becoming the opposite toward the end.

Delivery

Delivery may or may not be completion. Project completion often requires that the product function after delivery at a location the customer designates. Even if not explicitly the case, there may be an implied warranty that requires it. Thus, responsibility for delivered goods after they leave the contractor's facility is always an issue to be considered at project inception.

Consider the situation where the customer has witnessed final system tests at your company and accepts the system. After delivery to the customer's plant, the system does not perform adequately. (In the case of the Hubble Space Telescope, the performance inadequacy was not discovered until the telescope was tested in orbit, which was years after the primary mirror left the optical manufacturing plant.) What can the project manager do? As a minimum, you are going to have to participate in fact finding. At one extreme, the acceptance criteria will have been absolutely clear; the tests at your company will have been unambiguous; participation in and witnessing of these tests will have been done by qualified and responsible senior personnel; and the performance shortfall at the customer's facility will be attributable entirely to the customer's actions.

More likely, questions will arise about the acceptance test, the specific equipment used to perform it or what might have occurred to change the system during delivery. Regardless of contract form or your desire to obtain more business from that customer, the customer's final happiness with how well the system solves his or her problem will affect your reputation; so you may have to engage in a lot of extra, perhaps unpaid, effort to help get the system working. In some cases, you may be able to show the fault was theirs and get them to pay you for your extra effort.

Once the contractor has turned over possession of the goods, the contractor's control is greatly diminished. Therefore, it is important to have clearly stipulated conditions for acceptance, including payment terms.

Documentation Reports

It is not at all uncommon for a project to require the delivery of documentation as well as some other tangible output. Such documentation, in addition to a final report, might include a spare parts list, instruction manuals, and as-built drawings. Even the format may be specified.

Because furnishing these items may delay completion of the principal project output, they should be identified as a separate line item in the original contract. Thus, billing for the principal portion of the contract may accompany its delivery and billing for the documentation reports may follow later. Figure 22-3 illustrates a conservative time plan to assure that final documentation reports are delivered prior to the stipulated deadline. An analogous schedule could be constructed for hardware (although the shipment method might require shipment by a heavy duty freight carrier).

It is frequently difficult to complete documentation at the end of the project for two reasons: First, many technical specialists are poor writers or are reluctant to write. Second, in many instances, the people who have essential knowledge have long since been assigned to some other activity and are no longer working on the project. The second problem can be eliminated and the first problem diminished by what I call "incremental documentation." At the project's inception, you can prepare an outline of all final documentation and include this in your definition (or proposal) statement of work. Throughout the project, the responsible task manager or other specialist provides a few sentences or paragraphs when each key task is completed. Figure 22-4 illustrates this concept of incremental documentation, for which each task (the solid line) has a brief documentation appendage (the dotted line continuing after the solid line) during which the task manager completes a few sentences, paragraphs, or pages pertinent to the task just completed. These are inserted into the outline at the correct place. This is relatively painless and provides the essential information for final documentation later. These contributions might be provided via a local area network or on computer disks, thus reducing transcription requirements.

FIGURE 22-3. *A hypothetical conservative completion schedule to assure on-time delivery of required documentation.*

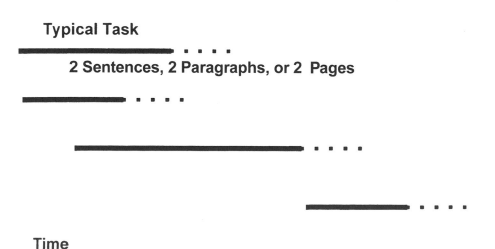

FIGURE 22-4. *The concept of incremental documentation.*

COMPLETION CONSEQUENCES

Project completion may be viewed as a boon or doom. The customer, the contracting organization, the project manager, and the project personnel may not all see it the same way.

Four Affected Parties

For the project manager, completion may be an opportunity for promotion, but many project personnel may find themselves laid off if there is no other work. If the project was badly managed, its manager may receive a less favorable assignment in the future, and personnel who did an outstanding job may have a choice future assignment. The contracting organization's view of project completion depends on customer approval and project profitability. The customer may be unhappy because he or she specified the wrong acceptance criteria but must admit the contractor has met the specifications.

Everyone does not have the same stake in completion, and the project manager must understand the differences.

Thus, there is no reason to assume that all four parties will have the same view of project completion. Project managers must realize that they may have a very different stake in ending the project than the other three parties. Consequently, the project manager must prod, cajole, or offer inducements to those for whom completion is not obviously desirable.

Completion consequences are also influenced by the reasons for termination. It is certainly best to end the project because all the objectives have been satisfactorily achieved. It is a bad situation if one or more dimensions of the Triple Constraint have been missed substantially.

Personnel Reassignment

Project completion requires reassignment of people. We have now come full circle. The (temporary) project is no longer imposed on the rest of the (permanent) organization. This frequently will necessitate a reorganization of the parent entity because the mix of remaining project work is such that the previously satisfactory organization is no longer appropriate.

The other crucial aspect of personnel reassignment is timing. If a person's next assignment is a choice one, he or she will normally be anxious to start and will lose interest in completing the present project. Conversely, if someone's next project assignment is undesirable, he or she may stall. When no assignment is obvious and layoff or termination is probable, personnel may even attempt sabotage to stretch out the present project assignment. Incentive bonuses for timely completion may be useful to counteract this potential problem.

A person's perception of what will happen when the project ends will affect his or her work as termination approaches.

The project manager can cope with these tendencies to some extent by selecting the time he or she informs project personnel of their next assignment. But if the contracting organization has a reputation for terminating personnel at the end of projects, there is little a project manager can do. The best situation is one in which all project personnel can count on their good work being recognized and appreciated and there being a selection of future assignments.

There are many options for personnel reassignment when projects end.

Even if no specific new project assignment is available when personnel need reassignment, there are still options. For example, personnel can write an unsolicited proposal, prepare an article for publication, work on an in-house development effort, or attend a short course or seminar. Temporary assignments such as these can be used constructively to fill in valleys in the project workload. They also can be used as a motivational tool if they are authorized so as to make participation a mark of recognition for a job well done.

Organizational Changes Due to Completion

When any project ends, for whatever reason, the organization is altered. There is a new body of knowledge in the organization. This is not just tangible information, but also new skills for many people. New working relationships have been established, both within and external to the organization, and these alter the informal organization, even if the formal organization is still unchanged.

INCREASING THE ODDS OF SUCCESS

There are both external and internal factors that influence how well a project satisfies its Triple Constraint.

External Factors

A knowledgeable customer, high priority, and clear objectives aid successful completion.

Customers for projects seem to be divided into two broad categories: knowledgeable and shortsighted. The shortsighted customers tend to emphasize the buyer versus seller relationship and to some extent create an adversary relationship between the two organizations. Conversely, knowledgeable customers realize that their stake in project success is ultimately just as great as that of the performing contract organization. Thus, a knowledgeable customer will become involved in the project in an effective, as opposed to destructive, manner. Such a customer will specify expected reviews and include them in the original job definition. Beyond this, he or she will attempt to ask the tough questions and to carry out probing reviews of the contractor's work, not to embarrass but to assure that all significant issues have been dealt with appropriately. Any required changes will be negotiated intelligently.

High priority projects inevitably seem to have better outcomes than lower priority projects because they tend to win all competitions for physical and human resources. This is not to say low priority projects lack top management support; top management clearly wants all projects to succeed, but the lower priority projects are at a relative disadvantage.

Clear and stable project objectives are a sine qua non of project success. Objectives can and do change during the course of many projects, but not on a daily or hourly basis. Thus, committing these objectives in writing helps fix them in everyone's mind. Revising them when they must occasionally change is also a requirement of success.

Internal Factors

A qualified, experienced, competent leader is vital, as is a balanced team. Having a team with a balance of skills and getting teamwork from it can be somewhat contradictory. People with very similar backgrounds tend to get along better; so it is easier to promote teamwork in a group composed entirely of, for instance, electrical engineers. Nevertheless, a successful project usually requires a team be composed of more than electrical engineers. Thus, the project manager is confronted with merging people with diverse backgrounds into an effective and harmonious team.

A good leader, a balanced team, the right-sized work packages, careful replanning, and orderly termination contribute to project success.

Having the properly sized work packages helps you avoid two potential problems. Complex, difficult work packages should not be assigned to junior people, who may be overwhelmed by them. Simple work packages should not be assigned to senior people, who will not be challenged by them.

Because projects will almost never be carried out exactly in accordance with the original plan, replanning is a constant requirement in project management. Project termination, especially the reassignment of personnel, requires active planning well before scheduled completion.

COMPUTER SOFTWARE

Figure 22-5 illustrates the concept of incremental documentation. Four documentation tasks (A*, B*, C*, and D*) have been inserted to follow the respective tasks, and all of these are shown as additional predecessors to the final report. (Each of the documentation tasks is shown as having one day's duration, although a much shorter elapsed time would normally be expected.)

292

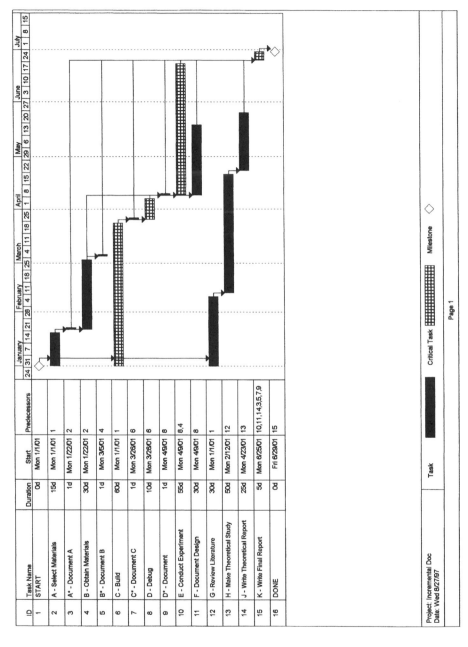

FIGURE 22-5. TSTETIL critical path schedule with four incremental documentation tasks added to the schedule plan of Figure 7-19.

TYPICAL PROBLEMS

Sometimes not only subordinate personnel but also the project manager must change during the project's life. The manager for the initial phases may be great at the inception but become stale with time or bored by routine wrap-up activities. The solution in this case is to change project managers, and both upper management and the project manager must be alert to this possibility, which then requires time for the replacement project manager to review (and perhaps revise) the plan so he or she is comfortable with it.

MATERIALS STUDY PROJECT

At the end of the materials study project, NERESCO could have asked why more materials had not been tested. In such a case, SUPROMAC must cite the contract and remind NERESCO that an amendment (more materials, more time, more money) could be negotiated to permit this.

Assuming NERESCO is thus satisfied with the number of materials actually tested, SUPROMAC can complete the materials study project by submitting its written report to NERESCO. However, even if that is the only contractual obligation, it may be better to present it as a part of an oral briefing in order to (1) improve communication by allowing some give-and-take discussion and (2) provide an opportunity to increase the customer's happiness with the results.

HIGHLIGHTS

- Although the project life cycle starts with the definition phase and ends with the completion phase, project completion and customer acceptance depend on agreements reached during the definition phase.
- Personnel needs may change throughout the project life cycle.
- It is best to end a project because all dimensions of the Triple Constraint have been satisfied.
- The project manager must realize that project completion may not be good for all involved parties and plan for an orderly end well in advance of its scheduled time.
- Both internal and external factors contribute to project success.

23

Final Wrap-Up

Projects end with their completion, but there frequently are postcompletion activities that are necessary to the project and that may be viewed as part of it. They are not, however, necessarily part of the basic contract. These are discussed in this chapter.

CONTINUING SERVICE AND SUPPORT

Continuing service and support may be an obligation. If they are, it must be understood who is to pay for them and when. This is often left to be negotiated when the project is completed. Negotiating at this time may be desirable because of an initial inability to see what may be involved, but it leaves a potential Pandora's box at the end of the project.

Some projects require postcompletion service and support.

The contractor should view continuing service and support as an opportunity and not merely as an obligation. His or her employees will be working with the customer's personnel, providing continuing service and support, if it is included in the project. In so doing, they will have informal opportunities to explore ideas with the customer's personnel and hear about real problems the customer is facing. Thus, these contacts provide the basis for future business opportunities.

OWNERSHIP RIGHTS

Outline patent rights and hiring policies in the contract.

Patents and special facilities required for contract performance have ownership value. In general, if it is a cost reimbursable contract in which the customer pays all the costs incurred, ownership rights revert to the customer at the end of the job. If it is a fixed price contract, ownership rights generally revert to the contractor unless otherwise stipulated in the initial contract. These can be points of contention unless they are discussed and clearly resolved in the initial contract. In any event, patent applications must be filed if any seem justified. The party of ultimate ownership must expect to pay for this activity.

There is also the issue of people. Obviously, neither the customer nor the contractor owns any people. But "no compete" clauses in recruiting customer personnel into the contractor organization or vice versa are often in the initial contract. If they were not, the customer might hire the contractor's personnel to perform the continuing service and support the contracting organization presumed it was going to perform and be reimbursed for.

AUDITS

There often are postcompletion audits, especially in contracts performed for governmental entities. Such audits may occur many years after the project work is otherwise complete.

It is therefore absolutely essential to retain records for the required duration and even more important to file and document them in an organized and thorough way. Many of the people who could explain some audited item may no longer be available when the audit is performed; so the contractor might lose an important claim if this is not done.

A project is not complete until the customer pays the bill.

Some portion of the final payment on a contract may be withheld until the audit is completed. Therefore, a final financial summary may not be possible until the audit is performed, and this can be months or even years after the other work is finished.

What can you do if your customer does not pay the bill? The responsibility for obtaining payment usually resides with the accounts receivable department and is thus outside the normal purview of the project manager. Nevertheless, an effective project manager may be able to help persuade a slow-paying customer to pay promptly. And the project manager, in considering the possibility of obtaining more work from the customer, will be concerned with the customer's payment history because that can influence the desirability of further work. If further work is desired, the project manager does not want the accounts receivable department to take hostile action against a delinquent account. Thus, the project manager and the accounts receivable department must discuss these issues and mutually decide on the mix of persuasion, cajoling, and legal action to take.

Regardless of whether the sponsor conducts an audit, every project manager should organize a project postmortem. These are also known as postcompletion or postimplementation evaluations or reviews or, perhaps more helpfully, as postproject learning. Some of the aspects to investigate and draw lessons from are the procedures that were used; any processes, tools, or methods employed if hardware was involved; the abilities of the organization and individual staff members; and any training that would be

Those who cannot remember the past are condemned to repeat it.

desirable. The goal is to record the "lessons learned." Because it is difficult to be objective, the postmortem might be conducted by an outsider or a person not intimately involved in the project. He or she should interview each of the key participants and provide a summary report. This report might be presented in a briefing session for the key project personnel before its final submission to the sponsoring organization. The goal of such a report is to help the organization avoid unnecessary mistakes in the future and point out how to revise procedures so that similar future projects can be carried out more effectively and expeditiously. In some cases, it is best to prepare for such a postmortem by having the outsider who will conduct it interview key project personnel at two or three important points during the project to capture their thinking at those points while it is still fresh.

PEOPLE ISSUES

After being assured that all project personnel have been reassigned (or laid off or terminated if necessary), the project manager has two other things to do. First, send personal letters of thanks, appreciation, or praise to project personnel. Second, send a brief wrap-up report to management. It is smart to cite your own successful performance in this.

TYPICAL PROBLEMS

Basically, the problem here is simply doing it all. The press or excitement of new items to do often leads to the omission of some postcompletion activities. Organizations frequently never get around to performing a formal postmortem, so the same mistakes are repeated in the subsequent projects. The solution is to recognize your responsibility in getting these things done.

MATERIALS STUDY PROJECT

In the case of the materials study project, SUPROMAC could use the final report to submit either an oral or a preliminary written proposal for further work. Assuming it is well received, they could continue to discuss this with NERESCO until a new project is defined and initiated.

HIGHLIGHTS

- Continuing service and support activities may lead to future business opportunities.
- Contracts often include assignments of patent rights and "no compete" personnel recruitment clauses.
- Records must be carefully kept in case there is a postcompletion audit.
- Managers should send letters of appreciation to project personnel and a wrap-up report to the boss.

Part 6

Other Issues in Project Management

24

Small Projects

Relative to large projects, small projects have both advantages and disadvantages. Being smaller, they are easier to understand and less likely to get into difficulty. But there are less time and money to recover from anything that goes wrong, and they inevitably lack high priority.

SIMPLIFIED MANAGEMENT

Figure 24-1 is a form that can be used to plan and monitor small projects. It shows the plan for the three research tasks of the materials study project (assuming, for the purpose of illustration, that this is a small task within the research group). The form allows the work breakdown items and the schedule to be entered in either a network diagram or a bar chart form. The overall plan for labor and nonlabor cost may also be entered. As actual costs are incurred, they can be entered and variances noted. In this particular form, the labor categories do not exactly match those previously used, but the form is merely illustrative and could be changed to include any categories. Note also that the form uses months for the time horizon across the top. This could be changed to weeks or any other convenient time frame. An alternative form, which lacks specific cost information, is illustrated in Figure 24-2. Using a spreadsheet, such as the one in Figure 9-2, is an alternative to filling out forms such as those in Figures 24-1 and 24-2.

Small projects have a Triple Constraint to plan and monitor.

You can also use computer-based project management software. I have used Microsoft Project 4.1 to illustrate the utility of computer-based project management software throughout this book and have commented on some improvements provided in Microsoft Project '98. Microsoft Project works well for both small and large projects. If it or other computer software is not otherwise justified or

PROJECT MANAGER _Tom Richards_
PROJECT NO. _T-84-03_ DATE _26 Dec_
PRIOR COST _—0—_
START: MO _Jan_ YEAR _____ TITLE _Theoretical Work – Materials Study Project_

MONTHS

KEY PROGRAM ELEMENTS	1	2	3	4	5	6	7	8	9	10	11	12
G – Literature Study												
H – Theoretical Study												
J – Final Report												

FINAL REPORT – Scheduled Submission

			1	2	3	4	5	6	7	8	9	10	11	12
Labor Hours	Senior Professional (Tom Richards)		120	160	200	160	120							
	Professional (Jr)				400	240								
	Technician													
	Total Plan													
	Actual													
Labor Cost, M Dollars (Incl. OH, Not G&A)		Plan	6	8	26	17.6	6							
		Actual												
Non-Labor Cost, M Dollars		Plan	1											
		Actual												
Outstanding Commitments, M Dollars														
Total Cost		Plan	7	8	26	17.6	6							
		Actual												
Cumulative Cost, Labor		Plan	6	14	40	57.6	63.6							
		Actual												
Cumulative Cost, Non-Labor		Plan	1	1	1	1	1							
		Actual												
Total Cumulative Cost		Plan	7	15	41	58.6	64.6							
		Actual												

LEGEND

○ TASK
△ DECISION POINT
☐ COMPLETION OR FALLOUT

Page _____1_____ of _____1_____

FIGURE 24-1. Simplified form for small project planning and monitoring.

required, forms such as those shown in Figures 24-1 and 24-2 or a spreadsheet may suffice.

PROBLEMS

There are four causes for the problems unique to small projects: tight schedules, tight budgets, small teams, and low priority.

Tight Schedules

A small project typically is planned to have a shorter schedule than a large project. Thus, the inevitable "getting up to speed" consumes a larger fraction of the available time. A one-week or one-month delay on a long program is less significant than the same delay on a two-, three-, or four-month program, a typical period for a small project. The implication is that small projects must be initiated at the very first opportunity and the project manager must devote a relatively large amount of effort to assuring schedule compliance.

Tight Budgets

A small project budget will be less than a large project budget. Hence, the absolute amount of money available for contingency must necessarily be less. When the inevitable revolting development occurs, there is less maneuvering room in which to cope with the consequences. The typical two-week lag in cost reports represents a significantly larger

Small projects lack time, money, personnel, and priority.

fraction of the overall project time and again leads to less reaction capability when some revolting development is discovered. In addition, the cost of time devoted to the program management function is likely to be a larger fraction of the total project budget. Thus, extremely close attention to cost is required on a weekly, if not on a daily, basis.

Small Teams

A large project can typically have the full-time attention of a functional expert (for instance, a thermodynamicist); a small project must make do with the part-time assignment of such specialists. Thus, the small project must compete against other projects for the specialists' time. In some cases, especially where computer programming is called for, this can be a major problem. Each time a person begins a particular programming task, he or she will spend a certain amount of time "getting up to speed." Thus, time is lost in reindoctrination. The problem may be worse with

PROJECT TASK PLAN AND SCHEDULE

DATE _____
PREPARED BY _____   REV. BY _____   PAGE _____ OF _____

PROJECT NUMBER _____
PROJECT TITLE _____
WORK UNIT/DESCRIPTION _____

YEAR

TASK SCHEDULE: PERIOD ENDING

TASK NO.	TASK DESCRIPTION	RESP.	DEPENDENCY TASKS	HOURS			DATE	
				PLAN	TO DATE	PLAN	PLAN	ACT.

Legend

△ SCHEDULE EVENT ONE TIME

△ RESCHEDULED EVENT NUMBER
INDICATES RESCHEDULING SEQUENCE

▲ COMPLETED EVENT

△—△ SCHEDULED EVENT TIME SPAN

▲—△ PROGRESS ALONG TIME SPAN

△—▶ CONTINUOUS ACTION

◇ ANTICIPATED SLIPPAGE

◆ ACTUAL SLIPPAGE (COMPLETED)

FIGURE 24-2. An alternative form for small project use.

computer programming, but it is not confined to that specialty. Confronted with this reality, the effective project manager should attempt to bargain for full days whenever part-time resources are required.

Low Priority

Studies have shown that high priority projects are more likely than low priority projects to be completed successfully because they will normally win each competition for any key resource. If you have a low priority project and another person has a high priority project and you both request the model shop make parts for your projects, the person with the high priority project will have his or her needs satisfied first. And it is unlikely that a small project will ever have the same significance for an organization as a large project, which means low priority is more common on small projects than on large projects. Finally, small projects have less visibility, therefore less chance for personal glory; so motivation can be less.

Imagine that you are spending half your time managing a small four-month project and spending the other half working on a much larger project. After two months you discover that the small project is running late and will now require three-quarters of your time to complete it on schedule. There are at least four options to consider:

1. Be late on the small project.
2. Request that you spend only one-quarter time on the large project so you can spend three-quarters of your time on the small one.
3. Request paid overtime approval for the small project.
4. Work unpaid overtime on the small project.

The choice among these, and any other viable options, can be aided by a decision tree or a qualitative matrix.

TYPICAL PROBLEMS

There is another, insidious problem, namely, that of projects starting small and escalating. Again, "staying on top" helps, as does switching to more extensive, formal project management techniques as the project grows.

Beware that small projects do not escalate.

MATERIALS STUDY PROJECT

As shown in Figure 24-1, Tom Richards is to perform the theoretical work on the materials study project (tasks G, H, and J). If he gets ill, quits, or is reassigned to a higher priority project, there is very little time in which to recover (that is, only the slack time until the project's final report, task K).

HIGHLIGHTS

- Four causes of the problems unique to small projects are tight schedule, tight budgets, small teams, and low priority.
- Another problem associated with small projects is a tendency for the project to grow.

25

New Product
Development Projects

New product development projects are prevalent, have some unique aspects, and are very important for many corporations. This chapter covers key characteristics of these projects, provides a framework with which to understand how the desired profits may be more quickly obtained, and stresses the importance of assuring that adequate resources are available.

WHY NEW PRODUCT DEVELOPMENT PROJECTS ARE UNIQUE

In common with small projects (see Chapter 24), new product development (NPD) projects deserve special attention in this book (this chapter). NPD projects also provide a unified way to demonstrate the practical role of many topics previously covered in this book. NPD is an omnipresent economic undertaking of great commercial importance worldwide, and project management tools and techniques are especially useful for the varied projects that arise in NPD. Although the phase "new product development" is commonly used, it embraces both tangible products (for example, a new drug, child's toy, commercial airliner, or frozen food entree) and less tangible services (for example, a new type of certificate of deposit, combination of credit card terms, or frequent user program).

NPD projects differ in six important ways from other commonly encountered projects.

1. The Triple Constraint dimensions usually have particular characteristics:

   - Competitive pressures, the commercial "window of opportunity," and the substantial advantages of being the first market entrant with a new product are such that the schedule is normally critical.

     Schedule is normally paramount for an NPD project.

   - The performance goal is also important, but it is

rarely an absolute standard. The performance goal is normally established as a marketing trade-off with the forecasted schedule availability and relative benefits the product is expected to provide to the prospective users. In general, a simple new product can be developed quickly, whereas a more complex or higher performing product requires more development time. The performance goal must also include the target selling price and gross margin, so the expected *factory* cost is important.

- The actual *development* cost per se is relatively unimportant after the financial payback has been calculated and justified initially. The critical issue is more commonly the allocation of limited resources (principally human, but also physical) among competing NPD investment opportunities. It is more important for senior executives to constantly ask whether the continued investment in a particular NPD project is the best (or, at least, justified) use of the firm's limited resources and not to focus on whether the NPD project is over or under budget.

2. They are self-funded by a company, rather than undertaken as a contract to deliver some result for another organization. (NPD obviously can be contracted by a sponsoring company to a specialized NPD company, but this is not a common way to develop new products. Most NPD projects are carried out within the sponsoring company, although there may well be subcontracts to deliver parts, components, subassemblies, or production equipment.) Other things being equal, lower project cost will be more profitable than higher. The latter frequently result when projects are undertaken on a contract basis.

3. NPD can involve many different kinds of projects. These include products that are new to both your company and the market, new to the market (product line extensions), and new to your company ("me too" products), minor variations on an existing product, sustaining efforts (typically technical assistance to the manufacturing operation), troubleshooting, and so on. These may be very small projects (a few person-hours or a few days) to very large (hundreds of person-years for a new commercial jet airliner, for instance). Some projects may require little technology (a revision to an operator's instruction manual); others may involve the complex integration of several technologies (for example, mechanics, optics, electronics, software, and reagent chemistry for a new medical diagnostic system).

4. Isolated, so-called "one off," NPD projects are common. Unfortunately, these are far less profitable than efforts that support an integrated family of new products based on a common platform. The latter provide numerous benefits to producers and can offer substantial benefits to buyers and users.

5. It is important to satisfy both a customer (the direct buyer) and a user (the person who actually employs the product). For many consumer goods, these are the same, but pet food and baby products are important categories for which the buyer and user are not the same. In most cases, there is also a distribution channel from the manufacturer to the point of sale, and the channel intermediaries (retail chains, stocking distributors, manufacturer's representa-

tives, and others) are critical factors in commercial success or failure. Market research is therefore a critical ongoing activity during development.

6. Timely and cost-effective NPD inherently requires a multifunctional project team that carries out many tasks simultaneously. Coordination is therefore crucial and can be greatly encouraged by the use of a TSTETIL (such as shown in Figure 13-3) where the nature of each function's involvement in each task is clearly indicated.

Multifunctional teamwork is crucial.

A GENERAL FRAMEWORK

Understanding some ways to cash in more quickly on your investment in an NPD project can help. The next section frames the issue, exploring what profit means and why a lack of clarity may make it elusive, and then defines four intervals of time from inception to the end of the product's life.

The Profit Objective

The widely practiced stage-gate new product development process is typically portrayed to start with an idea (the proverbial "lightbulb") and end with "launch" and then a "pot of gold" or a "bag of cash," as shown in Figure 25-1. In fact, even if you have successfully navigated through the "stages and gates" course, there is nothing automatic about (1) making enough money to justify the risky new product development effort or (2) making money quickly enough to yield a satisfactory discounted cash flow return.

Research has shown that the profit objective is normally multidimensional, having both project-level and firm-level dimensions or components. There are often two

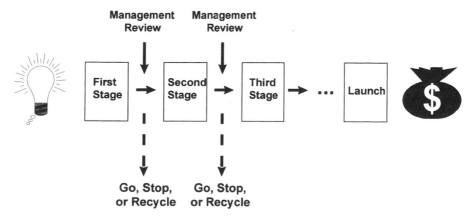

FIGURE 25-1. *A general stage-gate process.*

simultaneous project-level measures, which may be comprised of some combination

The profit objective is multidimensional.

of market share, revenue, profit, competitive advantage, and customer satisfaction or acceptance. The firm-level objectives can include both a financial and a performance measure, such as return on investment, percent of profits from new products in recent years, the success versus failure ratio, or a fit to business strategy.

Regrettably, many firms have installed a stage-gate NPD process, but the NPD project teams lack a clear understanding of either the firm's NPD strategy or the multidimensional profit objective. Lacking knowledge of the desired destination, an NPD project team depends on luck to be successful. I am aiming to (1) broaden the reader's view of NPD to encompass the entire process from inception to the realization of the profit objective (whatever that may be in your firm) and (2) identify a few ways to shorten the time from inception to profit.

Four Intervals (or Periods)

Figure 25-2 illustrates five events and four intervals, for which I use the abbreviations FFE (fuzzy front end), S & G (the traditional stages and gates portion of the process), SS (the postlaunch period until the time at which the profit objective is achieved),

There are three general ways to shorten the time to profit.

and CS (continued sales subsequent to achieving the profit objective). CS is "gravy" and will not be further considered.

Figure 25-3 shows that it is possible to shorten the time to profit by shortening the three initial intervals. Alternatively, it may be possible to lengthen the FFE to subsequently shorten S&G and/or SS or to lengthen S&G to shorten SS.

Fuzzy Front End Figure 25-4 shows that you can initiate the FFE with any of four stimuli. But all four of these factors (strategy, ideas, technology, and markets) must be carefully considered and satisfactorily resolved in this initial interval. The

Five Events

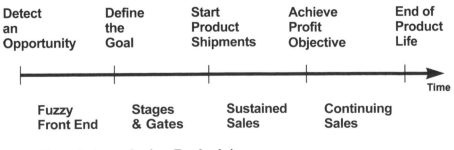

Four Intervals (or Periods)

FIGURE 25-2. Four intervals (or periods).

Nominal Plan

FFE   S&G   SS

Shorten Any or All Periods

Lengthen FFE to Shorten S&G and SS

Lengthen S&G to Shorten SS

FIGURE 25-3. Three options to shorten the time to profit.

Technology

Strategy ←→ Ideas

Markets

FIGURE 25-4. Four FFE factors.

goal of the formative period is to remove uncertainties because a development schedule is not credible if there are major unknowns about the market, the technology, or the production process. Failure to carefully execute the FFE is often the cause of extra, more costly effort during the subsequent S&G or SS intervals.

Project management tools are most useful in situations where what has to be done is known with reasonable certainty. They are less useful for the overall discovery research situation where each experimental result can dictate a new course of action; however, the tools are invaluable for managing each experiment to increase the likelihood that each experiment—once decided upon—is completed quickly and efficiently. To put this differently, the fuzzy front end and fundamental discovery research should be structured to run numerous knowledge-building experiments quickly.

As one example of a company that has developed a systematic approach to the FFE, Kodak previously started with an idea (generally from a technologist) and then went directly into their S&G development process. They now select an opportunity, then go through their "discovery and innovation" process (which entails forming an innovation charter; seeding ideation; building models, samples, and prototypes for screening; and developing a business plan). Only then do they initiate the S&G part of development.

> **Strategy, ideas, technology, and markets must be considered in the FFE.**

The model shown in Figure 25-4 is general and therefore potentially very useful. For example, either market or technology discontinuities can provide excellent entry points into the FFE. Senior executives may want to change the firm's strategy (but only thoughtfully and carefully) as a result of a new "killer" idea.

The "window of opportunity" for any effort must be considered early in the FFE and may lead to an early termination or require the acceleration of an effort. In all cases, the end of the FFE is a solid business case that is judged against understood criteria.

Stages and Gates   Figure 25-5 shows the efforts in S&G (set specification, time-critical new product development, and continuous learning) and the FFE and SS intervals that, respectively, precede and follow. The reason to show these other activities is that many of the resources required for S&G are also involved, and your firm's resource allocation (or lack thereof) has important implications for the timeliness of S&G.

> **The time to market depends on the difficulty of the specification and the effectiveness of the resources.**

Figure 25-6 (which is based on the concept illustrated in Figure 8-5) depicts the key relationships that affect speed during the S&G interval. This is the critical S&G trade-off. In general, a product with a more difficult specification normally has a longer time to market (that is, launch) than one with an easier specification. The time to market is crucial in some situations, such as in the $1 billion or greater investment for a new generation of a dynamic random access memory chip where the vast bulk of profits is earned in the first year of sales.

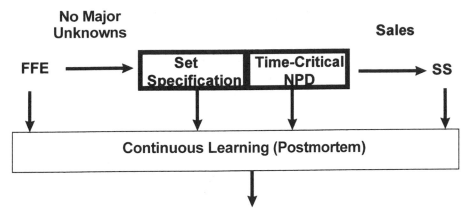

FIGURE 25-5. *Stages and gates.*

However, at any level of specification difficulty, the time to market can be shortened by making use of more effective resources; the converse is also true, that use of less effective resources causes a longer time to market. Thus, to shorten the time to market, you must aim for simpler specifications or plan to utilize more effective resources (either people or equipment). Here are three ideas to shorten the NPD interval:

FIGURE 25-6. *Options to accelerating the time to profit.*

1. *Clarify the specification.* Specifications can involve both objective (quantitative) and subjective (qualitative) elements, as discussed in Chapter 3 and illustrated in Table 3-1. The obvious problem is how to decide when this specification is met. Unless the means by which a subjective specification will be judged is clearly delineated, developers end up working at cross-purposes, and time is lost. When a subjective specification element is required, it is crucial to clarify in advance how compliance will be judged. In a small private company, the owner typically makes the decision, so he or she must be kept informed. In a large public company, it can be harder to identify a single decision maker, but this is what you must try to do.

2. *Capitalize on a product family.* Another way to simplify specifications is to plan to develop a product family (that is, use planned releases to spread the related new products over a reasonable time span). This approach offers seven potential benefits:

   You can start quickly with a relatively simple member.

   Early entry provides real market data.

   Lower costs are possible due to common parts and processes.

   There are fewer risks because portions of the product are already proven.

   Follow-on members can reach the market quickly.

   It is easier to defer "feature creep."

   You can provide a "hook" to extend product life (especially for software upgrades).

   There are numerous examples of product families in the consumer electronics market. Go into any mass merchandise outlet (such as Circuit City) and look at the wide variety of products with slightly different features and prices being offered under a single brand (for example, Sony tuners or CD players). Hewlett-Packard has replicated this tactic with its ink-jet printers and Pavilion personal computer family. In the case of the latter, the initial market introduction offered buyers seven models.

3. *Use flexible gates.* Flexible gates, Figure 25-7, in a stage-gate process can shorten the time to market during the S&G interval. However, replacing rigid gates with permeable and permissive gates increases risk somewhat. Nevertheless, on balance, using flexible gates is often the lesser evil.

Problems after market launch can delay achievement of the profit objective.

Sustained Sales Problems The list of problems that can plague a new product when it reaches the market is almost endless, and, sadly, new types of problems continue to emerge to frustrate prospective users. No list, therefore, can be exhaustive, but the following common problems are illustrative:

Users encounter unanticipated difficulty with the product, such as arcane instructions, application incompatibility, or inadequate performance.

The producer or, worse, the user is saddled with high maintenance costs.

FIGURE 25-7. Gate options.

The producer or someone in the supply chain is forced to maintain a large spare parts inventory.

The producer incurs an unexpected product warranty cost and, perhaps, must cope with returned goods.

The producer discovers that the price-to-cost ratio is less than planned.

There are actions that can and should be taken in the FFE and S&G intervals to avoid SS problems and, thus, shorten the time to profit. Again, the list could be almost endless, so a few illustrations must suffice:

Design the product for ease of use.

Invest in really excellent user manuals.

Maintain a well-staffed help line.

Spend enough time to assure that the design is robust and satisfies all advertised performance specifications.

Plan for rapid product distribution after launch.

Facilitate the efforts of the sales force by providing helpful selling tools, conducting an active promotion and advertising campaign, and otherwise stimulating their motivation.

RESOURCE OVERLOADING

The key to assuring a short time to profit is the proper allocation of a company's limited resources. Both resource capacity and flexibility are limited, no matter how large the organization. In the case of baseball, we can ask how many great hitters

██████████████

Resources are limited.

██████████████

make great pitchers or how many catchers make great short-stops. In the world of new product development, not many highly trained specialists are flexible generalists; so a mechanical engineer who is available may not have the skills needed for a particular development project. Or a critically required test facility may not have the capacity to simultaneously accommodate two or more development projects.

The Resource Allocation Problem

In most companies, there are too many NPD projects under way at any given time, so all are starved for resources and move forward at less than the fastest speed possible. Worse yet, any problem with an existing product threatens immediate revenue and becomes a current crisis. This normally preempts development resources and thus interrupts some or all NPD efforts. Not uncommonly, longer range development projects are sacrificed for those of shorter range.

Two Solutions to Consider

Let's briefly review the use of project management software and project priorities, which are two means at least to reduce management's propensity to make unrealistic commitments of limited resources:

1. *Project management software.* As shown in Chapter 10, project management software can be used to identify resource conflicts before they occur. It can also be used to encourage multifunctional teamwork by serving as both a coordination and responsibility matrix. As illustrated in Figure 13-3, columns for each organizational function can be used for each task to indicate the nature of its responsibility in that task as accountable (A), concurrence required (C), or input required (I).

2. *Project priorities.* The benefit of establishing project priorities and communicating them is that development resources can be applied first to the highest priority effort. If the highest priority project never has to wait for resources it can productively utilize, it can be completed faster. Thus, the resulting product can begin to earn income, and resources can be redeployed. These resources can then be applied to the second priority project, and so on. If a company has enough resources to support several development efforts simultaneously, it can do so; but even in such a situation, the practice of concentrating resources on a smaller number of efforts so they move faster still provides an overwhelming advantage.

Resource Allocation Strategy

Every company, even the very largest, has limited human and physical resources. Some of these resources must be devoted to managing the existing products and markets and dealing with the day-to-day distractions that consume time (for example, staff meetings, holiday and other social events, personnel issues, and so on). Too frequently, NPD project resources consist solely of those that are left over. In most companies, however, these resources are not thoughtfully allocated among the following:

Minor product refinements, substantially new products, and major break-through efforts

The FFE, S&G, and SS intervals

High and low risk efforts

Near- and long-term projects

The growing literature on portfolio management explores this in more detail, but there is no magic formula. Resource allocation is *the issue* that top executives must actively manage: The use of project management software can elucidate specific potential resource shortages; thereafter, specific project priorities must be assigned to help resolve or overcome these shortages. Sadly, many executives do not recognize this need. As one NPD project manager said to me, "In what ways would it be best to communicate (and stress the need) to our upper management of the importance of establishing project priorities?" In other cases, the priority decision is left to the marketing department, but an NPD project manager complained, "How can we set up a project priority process in which marketing considers input from all functional groups to ensure that the priorities realistically fit the limited human resources?"

> **Top management must allocate resources to the most important NPD efforts.**

TYPICAL PROBLEMS

The initial specification and development plan for an NPD project commonly results in a time to market that is longer than acceptable, as illustrated in Figure 25-8. It is easy to invoke arbitrary reductions of time, but these will not actually occur without a change of plan. Figure 25-8 is a disguised real case in which the initial time to market was about three years but market competition required it be one year or less. In this case, the shorter goal was achieved by doing the following:

Putting several serial tasks in parallel

Simplifying the specification

Finding and using more effective resources

Using staged market introductions in a "family" plan of releases

(continued)

FIGURE 25-8. *An example of shortening the time to market.*

Premature promises by the members of the marketing or sales functions create another common problem for NPD project managers. In some cases, these functions optimistically assure prospects that certain specifications or delivery dates can be met. In the world of mass market computer software, the term "vaporware" has been coined to characterize early product announcements. In many cases, the claims have not been checked with the development or manufacturing functions. This can lead to a lack of credibility in the marketplace or to an excessive and stressful workload in the departments responsible for fulfilling these commitments.

HIGHLIGHTS

- Clarify and agree on the profit goals that are the objective.
- Manage all three initial intervals (FFE, S&G, and SS) to minimize the overall time to profit.
- Allocate limited resources to accelerate the highest priority efforts.

26

Computer Software

Chapters 6–10 illustrate how computer project management software can help you develop a project's work breakdown structure, time schedule, cost budget, and resource requirements. This chapter puts these substantial benefits into a more general perspective. Despite the software's utility, it is not a panacea.

WHEN AND WHERE TO USE COMPUTER PROJECT MANAGEMENT SOFTWARE

I have provided several illustrations earlier in the book with examples of computer project management software. The location of these illustrations reveals where I feel this software can be most useful. As shown in Table 26-1, an overview of management activities and Triple Constraint dimensions indicating where computer-based project management software is most likely to be applicable, the big payoff is in planning (or replanning) the schedule and cost dimensions. If the software you are using has a WBS outlining capability, it inherently helps plan the work to accomplish the performance goal, even though it will not assure successful completion of that work. The software provides the project manager with a "reality test," which forces him or her to think through the project in sufficient detail to devise some way to complete it. This may not be the optimal plan, but it is a starting point.

If you take the trouble to update the plan with actual schedule information, the software may also be helpful for monitoring. However, as Table 26-1 suggests, it will not help you explicitly with the definition or completion activities.

Table 26-1 has question marks next to the leading activity under the schedule dimension. If you are already facile with software and are performing many "what-if" trade-offs of schedule, then computer project management software can save you time, which you can use to interact with people on the project and thus be a more effective leader. Conversely—and this is a substantial trap for the unwary technical specialist—if you sit at a computer terminal performing the "what-if" trade-offs instead of spending sufficient time interacting with the human re-

TABLE 26-1 Best Use of Computer-Based Project Management Software When Actual Cost Data Must Be Manually Compared with Planned Cost (Top Lines) and Can Be Automatically Transferred into the Computer for Comparison with Planned Cost (Second Lines)

	Performance	Schedule	Cost
(Re)planning	+	++	++
	+	++	++
Leading		?	--
		?	?
Monitoring		+?	----
		+?	+?

sources on your project, the software will impede your ability to carry out the leading activity.

Table 26-1 shows different judgments for cost dimension utility of computer project management software. The top lines assume that the corporation's actual cost data are not automatically linked from the corporate cost-collection computer to the computer on which you are using the project management software. This is a common situation, especially in firms that are not in the business of selling project work per se. Examples include firms practicing project management to carry out

Beware of ineffective use of computer software.

new product development projects, minor facility rearrangements, reengineering, and so on. Because the tracking of cost is relatively unimportant in such situations, collecting the actual cost data can be very time-consuming, and they may be out of date by the time you can compare them to the plan. Time spent pursuing this is actually harmful to managing a project, especially in regard to leading and monitoring activities.

But if your corporate cost data are collected in a computer system that can automatically link them to your computer running the project management software (as illustrated in Figure 19-14), the cost dimension situation is identical to the schedule dimension situation as shown in the second lines of Table 26-1. Microsoft Project '98 (but not 4.1) has this capability if the cost data are available in suitable formats.

To reiterate what I said earlier in this book, using computer project management software is not the same thing as being an effective project manager. Nevertheless, this software can help a good project manager do a better job.

Project management entails much more than using project management software.

Conversely, insisting that a poor project manager use project management software will not make him or her a good project manager. Similarly, insisting that people working on a troubled project use the software will not fix the project's problems.

Figure 26-1 suggests that this software is adequate for many types of projects. I have not put numbers on either dimension in Figure 26-1 because hardware and software capabilities are changing rapidly. However, this software is certainly adequate today for a wide variety of projects. Because of increasing software and hardware capabili-

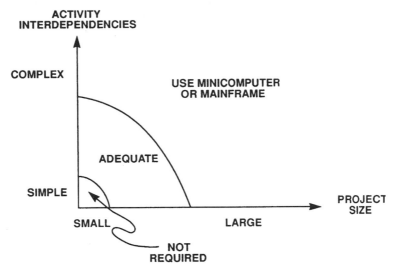

FIGURE 26-1. *Computer-based project management software is useful for a wide variety of projects.*

This material originally appeared in a videotape presentation produced by the Association for Media-based Continuing Education for Engineers, Atlanta, Georgia.

ties, newer versions will be able to handle even more complicated and larger projects in the future. Unfortunately, these improved capabilities often are obtained only with the trade-off of greater complexity. As I said in Chapter 24, software normally is not required for very small or simple projects.

Where the tasks on your projects, or parts of them, are repetitive, you can prepare a template and adjust it as needed only for the impact of different resources that will be applied. Thus, a standard task might take longer if an inexperienced person had to manage it.

CAUTIONS

All the computer software packages differ, so there is no uniform set of symbols or standards. Not even the terminology is standard. For example, "time-based PERT" or "time-phased PERT" is typically some form of precedence diagram but frequently lacks a linear time scale.

All the software packages have constrained formats, so you may never get exactly the displays you would most like to obtain. It may be a case of using the least objectionable package.

If your project requires major replanning of tasks to complete the work, you have the choice of modifying the original plan or creating an entirely new one for the balance of the work. In many cases, it is easier to do the latter because it may be very time-consuming to find and break the links to tasks that are no longer required.

**Be prepared to train
your key personnel
in software use.**

As project manager, you must assume that your key project personnel do not understand the output diagrams, reports, or formats. Consequently, you must plan to take time to train these people in all this material if you wish them to use it effectively. If your organization does not use a single standard package, you may have to teach your managers what the reports and symbols mean. The same is true for outside organizations that are involved. In some cases, an external customer may specify which software package you must use.

**You cannot manage
the team with
software alone.**

Much of project management's challenge arises because people are involved. Computer software is a mechanical device that will not solve the people problems. Figure 26-2 illustrates that a project manager cannot sit at a computer terminal, get output, publish it, and expect the team to comply. But if that output is used for discussion with team members and is revised (perhaps several times), there may be team "buy-in." The key to using computer project management software successfully is to have the team believe that the output represents *their* thinking and for them to understand it. One approach is to use an optical projector to exhibit the computer monitor's display on a large screen for team discussion during meetings of the key project personnel.

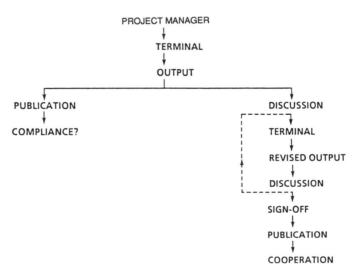

FIGURE 26-2. The output from computer-based project management software must represent a team consensus if it is to be effective.

This material originally appeared in a videotape presentation produced by the Association for Media-based Continuing Education for Engineers, Atlanta, Georgia.

Finally, as with all software, remember the aphorism "garbage in, garbage out" (GIGO). Unless you take the time to enter data that are as accurate and meaningful as possible, your output may be worthless.

Always remember "garbage in, garbage out."

TYPICAL PROBLEMS

The correct way to adopt any software is to first specify the problem you are trying to solve with it, second locate the software that solves the problem, and third find hardware that runs the selected software. In the case of many people who procure computer project management software, this process is reversed. They have existing hardware, they find some project management software that runs on it, and then they wonder why this does not solve their problem. Your challenge—and your opportunity—is to find and use computer project management software with which *you* can better manage *your* projects.

The other major problem is believing that the use of this software is synonymous with managing a project. It is nothing more than a very useful tool to assist with some aspects, and its use must not preclude the project manager from spending time working with people.

HIGHLIGHTS

- Computer project management software is readily available to assist a project manager in many important parts of the job.
- The use of this software will require time to train personnel.
- This software will not help the project manager to solve many of the interpersonal problems that occur.

27

Where Do You Go from Here?

In this chapter, I wrap up my discussion of the project management process with a summary of the key points you should remember as you put the tools and techniques discussed in this book into practice now or on your next project. Then I mention some trends I feel will impact how you can go about managing projects in the future. Finally, I list some sources for continuing your development of project management skills.

SUMMARY

The following list summarizes the five managerial steps for successful project management:

1. Define
   - Specific, measurable, attainable goals
   - Clear Triple Constraint priorities
2. Plan
   - Golden Rule
   - Three dimensions
         Work breakdown structure
         Time-based critical path schedule
         Task budgets
   - Contingency
   - Computer software
3. Lead
   - Understand own behavior

- Seek other people's strengths
- Anticipate interpersonal conflict
- Negotiate work packages
    Big WBS tasks for senior people, small ones for juniors
    Golden Rule
- Communicate thoroughly
4. Monitor
    - Project reviews
        Get and understand reports
        Ask questions, especially nondirective ones
        Task complete or not
        Try to avoid impending problems
    - MBWA
    - Change resources
5. Complete
    - Unambiguous specification
    - Incremental documentation
    - Reassign and thank entire team
    - Carry out postcompletion work

When you start a new project, you should review this list. Then turn to the chapter highlights, where appropriate, for a more detailed refresher. When necessary, you can then review portions of this book. Keep in mind that many of the topics are interrelated and cannot be dealt with in isolation, as illustrated in Figure 27-1.

SPECIAL INTERESTS

I have tried in this book to deal with project management in general. In some cases, I have cited examples applicable to a specific type of project, for instance, construction, engineering, aerospace, software, and so on. There are special issues in many of these types of projects. If you are working on a project for the Department of Defense, Department of Energy, or National Aeronautics and Space Administration, you may have to deal with C/SCSC, an extremely detailed specialized performance measuring system.

The heavy construction industry has various standards (steel sizing, concrete slump tests, plumbing and electrical wiring standards, and so on) you must understand and satisfy. Such standards are commonly available in construction handbooks.

Similarly, computer programming projects are best carried out using structured analysis and structured design (SASD). There are various tools to help carry out

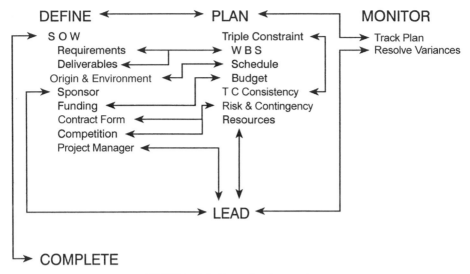

FIGURE 27-1. *Some key interactions.*

such projects; consider using them if project personnel know how to employ them and they are available.

New products for commercial and consumer markets today require increasingly shorter development cycles. For these projects, as discussed in Chapter 25, the overriding question in many cases is how fast they can be completed. As a result, it is often crucial to determine how to reduce the performance requirements to the minimum that will be useful in the market to permit you to get an initial product out quickly. More elaborate variations can follow.

In the case of other special interests, different issues might govern what is emphasized or how you go about the project. But my goal in this book has been to treat the subject generally and provide a selection of tools with which to attack any project. The differences between various types of projects are not as great as many specialists suggest, and there is a wide common ground. Small projects, as discussed in Chapter 24, can often be managed with simple tools.

OTHER SOURCES OF HELP

Sometimes you may still find that you are "over your head." In such a case, you may wish to retain a management consultant for assistance. The organization that certifies individual management consultants is

The Institute of Management Consultants, Inc.
521 Fifth Avenue (35th floor)
New York, NY 10175-3598

TEL: (212) 697-8262 or (800) 221-2557
FAX: (212) 949-6571
http://www.imcusa.org

There are other qualified management consultants unaffiliated with this organization. However, you have a better assurance of professional and ethical assistance if you choose a properly certified consultant.

You may also seek sources of more specialized assistance from two other organizations:

Product Development & Management Association
401 North Michigan Avenue
Suite 2400
Chicago, IL 60611-4267
   TEL: (312) 527-6644 or (800) 232-5241
   FAX: (312) 527-6729
   http://www.pdma.org

Project Management Institute
130 South State Road
Upper Darby, PA 19082
   TEL: (610) 734-3330
   FAX: (610) 734-3266
   http://www.pmi.org

THE FUTURE OF PROJECT MANAGEMENT

I think it is likely that more project managers will begin to have real-time project management status reports (technical, task status, labor hours, costs, and others) during the next decade. These reports may be provided on suitable terminals connected to mainframe computers or they may come from computer networks. Such real-time status information should allow project managers to recognize problems more quickly, which can reduce lost time if the project manager acts promptly.

In addition to real-time information, better graphic output devices (such as laser and color printers) will make it more convenient to keep network diagrams current. These diagrams can be distributed electronically and be produced at remote work sites. Better software will also facilitate graphical representation of management data that is now presented numerically, thus highlighting areas of change.

Finally, I think it would be desirable (but less likely) for organizations to install an adjudicator function. This adjudicator would hear disputes about internal agreements, such as task work orders, and render a decision. This would institute inside the organization an analog to the external use of courts or arbitration to resolve contract disputes. This kind of mechanism could go a long way to defusing internal corporate politics and the acrimony that accompanies it.

CONTINUING PROJECT MANAGEMENT SKILL DEVELOPMENT

Reading this or any other book will not make you an expert project manager. It takes time and practice for the skills to become second nature—and you will have to develop your own style, consistent with your skills, interests, and personality. There is no substitute for your own experience. As one project manager said to me:

> I want my subordinate to manage a complex study program, which has many technical variables. Furthermore, he must get support from other technical people and deal with their personalities. My subordinate is intelligent and analytical, but so far has been unable to put together a workable project plan. He has some difficulty in working with other people, but I believe he has the analytical tools for his job.

This illustrates that what works for one person may not be useful or appropriate for another. Thus, you should continue to experiment, read, and seek out other sources for continuing education.

Reading

There are numerous professional books on project management, specific aspects of project management, specific software packages for project management, leadership, planning, and other related topics. These usually have a short life in the turbulent market for books (as I can personally confirm by trying to find current editions of books I have cited in previous editions and other books). Thus, citations of specific books for further reading are likely to steer you into a fruitless search for books that are not available when you seek them out. Further, only a few bookstores in major metropolitan areas have a large selection of professional books.

Current professional books are most easily sought out by use of the Internet. I have had good success using amazon.com, which often provides some brief information about the book (and will accept returns if you are not satisfied with any book you have ordered). Other on-line bookstores probably can provide a similar selection and service.

In addition, you may want to read the *Project Management Journal,* published by the Project Management Institute. This is the only journal devoted to project management, but unfortunately it is mostly dedicated to heavy construction projects.

Continuing Education

Many universities and commercial organization sponsor "live" project management training, such as seminars of one day to one week duration. These vary in quality, teaching method, and subject matter; so you should determine who will lecture and lead these training programs. Then try to check out references by talking with prior participants. I find one-day seminars superficial (with the privately sponsored, commercial versions consisting largely of a "hard sell" for the sponsors' publications). The five-day short course versions may require more time (and money) than is necessary, so the two- or three-day versions are most appropriate for your consideration.

A six-hour videocassette course that is based on and augments this book is available from:

Instructional Television Network
University of Southern California
3650 McClintock Avenue (OHE 108)
Los Angeles, CA 90089-1455
TEL: (213) 740-0119
FAX: (213) 749-3289

The unique value of attending seminars or courses is the interaction with other project managers. No amount of reading or passive observation of videocassette courses—even if the materials are outstanding—can give you real practice with the development of human relationship skills, which are crucial when you must work with people.

If you cannot attend a university or commercial seminar, you may want to explore having such a seminar conducted in your own organization. Although you will not obtain the stimulus of interacting with personnel from other, different organizations, you will have a seminar customized to your own specific situation (perhaps using your own forms as examples), and usually such an "in-house" seminar can be much more cost-effective than sending many people to other seminars.

A FINAL THOUGHT

Good luck! May all your projects be successful.

Appendix 1

Abbreviations Used in Project Management

ACO—Administrative Contracting Officer
ACWP—Actual Cost of Work Performed
ADM—Arrow Diagraming Method
AIN—Activity-in-Node
AOA—Activity-on-Arrow
AON—Activity-on-Node
B & P—Bid and Proposal
BAC—Budget at Completion
BCWP—Budgeted Cost of Work Performed
BCWS—Budgeted Cost of Work Scheduled
CCB—Change (or Configuration) Control Board
CCDR—Contractor Cost Data Reporting
CCN—Contract Change Notice
CDR—Critical Design Review
CDRL—Contract Data Requirements List
CFE—Customer Furnished Equipment
CFSR—Contract Funds Status Report
CMO—Contract Management Office
CPFF—Cost Plus Fixed Fee
CPIF—Cost Plus Incentive Fee
CPM—Critical Path Method
C/SCSC—Cost/Schedule Control System Criteria
C/SSR—Cost/Schedule Status Report
CS—Continued Sales
CWBS—Contract Work Breakdown Structure
DAR—Defense Acquisition Regulations

DDR—Detailed Design Review
EAC—Estimate at Completion
ECN—Engineering Change Notice
ECP—Engineering Change Proposal
EF—Earliest Finish
EIN—Event-in-Node
ES—Earliest Start
ES & H—Environmental, Safety, & Health
ETC—Estimate to Complete
EV—Earned Value
FARS—Federal Acquisition Regulations
FCA—Functional Configuration Audit
FFE—Fuzzy Front End
FFP—Firm Fixed Price
FP—Fixed Price
FQR—Formal Qualification Review
FQT—Formal Qualification Testing
G & A—General and Administrative
GFE—Government Furnished Equipment
GIGO—Garbage In, Garbage Out
IR & D—Internal Research and Development
LAN—Local Area Network
LF—Latest Finish
LS—Latest Start
MBC—Management by Commitment
MBO—Management by Objectives
MBR—Management by Results
MBWA—Management by Walking Around
NPD—New Product Development
ODC—Other Direct Cost
PBS—Project Breakdown Structure
PCA—Physical Configuration Audit
PDM—Precedence Diagraming Method
PDR—Preliminary Design Review
PERT—Program Evaluation and Review Technique
PM—Project (or Program) Manager (or Management)
PMS—Project Management Systems
PO—Purchase Order
PR—Purchase Requisition
PWA—Project Work Authorization
QA—Quality Assurance
QC—Quality Control
QFD—Quality Function Deployment
R & D—Research & Development
R, D & E—Research, Development & Engineering

RFP—Request for Proposal
RFQ—Request for Quotation
S & G—Stages and Gates
SASD— Structured Analysis and Structured Design
SOW—Statement of Work
SS—Sustained Sales
T & M—Time and Material
TBAOA—Time-Based Activity-on-Arrow
TSTETIL—Time-Scaled Tasks with Explicit Taste Interdependency Linkages
WBS—Work Breakdown Structure
WO—Work Order
WU—Work Unit
3D—Three-Dimensional

Appendix 2

Glossary of Project Management Terms

Activity. A single task within a project

Actual cost of work performed (ACWP). A term in the Cost/Schedule Control System Criteria (C/SCSC) system for the costs actually incurred and recorded in accomplishing the work performed within a given time period

Arrow diagraming method. A type of network diagram in which the activities are labeled on the arrows

Bar chart. A scheduling tool (also called a Gantt chart) in which the time span of each activity is shown as a horizontal line, the ends of which correspond to the start and finish of the activity as indicated by a date line at the bottom of the chart

Bid/no bid decision. The decision whether or not to submit a proposal in response to a request for proposal

Bottom up cost estimating. The approach to making a cost estimate or plan in which detailed estimates are made for every task shown in the work breakdown structure and summed to provide a total cost estimate or plan for the project

Budgeted cost of work performed (BCWP). A term in the Cost/Schedule Control System Criteria (C/SCSC) system for the sum of the budgets for completed work packages and completed portions of open work packages, plus the appropriate portion of the budgets for level of effort and apportioned effort

Budgeted cost of work scheduled (BCWS). A term in the Cost/Schedule Control System Criteria (C/SCSC) system for the sum of budgets for all work packages, planning packages and similar items scheduled to be accomplished (including in-process work packages), plus the amount of level of effort and apportioned effort scheduled to be accomplished within a given time period.

Burst node. In a network diagram, a node at which two or more activities commence after the completion of the preceding activity

Buy-in. The process of making a cost bid in a proposal that is unduly optimistic or even actually less than the estimated costs for the project, which is done for the purpose of winning the job

Chart room. A room filled with planning documents displayed as charts, typically hung on the walls of the room, used on large projects, and usually marked to indicate current status

Commitment. An obligation to pay money at some future time, such as a purchase order or travel authorization, which represents a charge to a project budget even though not yet actually paid

Contingency. An amount of design margin, time, or money inserted into the corresponding plan as a safety factor to accommodate unexpected and presently unknown occurrences that judgment suggests will occur during the project (also called reserve)

Costed work breakdown structure. A work breakdown structure in which the costs corresponding to major elements of the WBS are also shown on the WBS

Cost plus fixed fee (CPFF) contract. A form of contractual arrangement in which the customer agrees to reimburse the contractor's actual costs, regardless of amount, and in addition pay a negotiated fee independent of the amount of the actual costs

Cost plus incentive fee (CPIF) contract. A form of contractual arrangement similar to CPFF except that the fee is not preset or fixed but rather depends on some specified result, such as timely delivery

Cost/schedule control system criteria (C/SCSC). A planning and control reporting system devised by the Department of Defense for its contractors to use, intended to foster greater uniformity as well as provide early insight to impending schedule or budget overruns

Critical path. In a network diagram, the longest path from start to finish or the path without any slack, thus, the path corresponding to the shortest time in which the project can be completed

Customer furnished equipment (CFE). Equipment provided to the contractor doing the project by the customer for the project and typically specified in the contract

Documentation. Any kind of written report, including such items as final reports, spare parts lists, instruction manuals, test plans, and similar project information

Dummy activity. An activity in a network diagram that requires no work, signifying a precedence condition only

Earliest finish. In a network diagram schedule, the earliest time at which an activity can be completed

Earliest start. In a network diagram schedule, the earliest time at which an activity can be started

Event. The end of any activity or task

Firm fixed price (FFP) contract. A contractual form in which the price and fee are predetermined and not dependent on cost

Fixed price (FP) contract. Same as firm fixed price

Float. Same as slack time

Functional organization. The form of organization in which all people with a particular kind of skill (such as engineering) are grouped in a common department, reporting to a single manager for that particular functional specialty

Government furnished equipment (GFE). The same as customer furnished equipment in projects where the customer is a governmental entity

Hardware project. A project in which the principal deliverable item is a product or functioning device of some sort

Latest finish. In a network diagram schedule, the latest time at which an activity can be finished

Latest start. In a network diagram schedule, the latest time at which an activity can be started

Matrix organization. The form of organization in which there is a project management functional specialty as well as other functional specialties and where the project management function has responsibility for accomplishing the project work by drawing upon the other functional specialties as required

Merge node. In a network diagram, a node at which two or more activities precede the start of the subsequent activity

Milestone. A major event in a project, typically one requiring the customer to approve further work

Network diagram. A scheduling tool in which activities or events are displayed as arrows and nodes in which the logical precedence conditions between the activities or events are shown

Periodic review. Any kind of project review conducted on a periodic basis, most commonly a monthly project review

"The Plan." A document or group of documents that constitutes all the plans for the project, frequently contained in a notebook or series of notebooks

Planning matrix. A matrix in which planned activities are listed on one side (usually the left) and involved people or groups are listed across a perpendicular side (usually the top) and where involvement of a particular individual or group in a particular activity is signified by a tic mark where the row and column intersect

Precedence diagraming method. A type of network diagram in which the events and activities are labeled in the nodes or boxes

Program. Used interchangeably with "project," as in "program management" or "program manager"

Program Evaluation and Review Technique (PERT). The form of network diagram in which events are displayed as nodes and connecting arrows indicate the precedence constraints

Progress payments. Payments made to the contractor by the customer during the course of the project, rather than at the end of the project, the terms of which are specified in the contract

Project. An organized undertaking utilizing human and physical resources, done once, to accomplish a specific goal, which is normally defined by a Triple Constraint

Project cost accounting system. A cost accounting system that accumulates actual costs for projects in such a way that total costs for all work in an organization can be allocated to the appropriate projects, normally providing monthly cost summaries; also used in cost planning to summarize the detailed task cost estimates

Project organization. The form of organization in which all or nearly all the people working on a project report to the project manager

Project plan. The entire plan for a project, consisting of the work breakdown structure, network diagram, and task budgets, but sometimes taken to mean only the network diagram

Project team. A term used in this book to designate the personnel working on a project who report to the project manager administratively, not merely for the work on the project

Proposal. A document (sometimes accompanied by models) that an organization submits to a prospective customer that describes work the organization is offering to do

Request for proposal (RFP). A document issued by one organization to another organization (or to several other organizations) describing work that the issuer wishes to have undertaken by the recipient(s) and inviting the recipient(s) to respond with a proposal

Request for quotation (RFQ). Similar to a request for proposal, except that the desired items to be procured are stock or catalog items, and only price and delivery time need be proposed

Slack time. In a network diagram, the amount of time on any path other than the critical path that is the difference between the time to a common node on the critical path and the other path

Software project. A project in which the principal deliverable item is a report or other form of documentation, such as a computer program

Statement of work (SOW). That portion of a proposal or the resulting contract that states exactly what will be delivered and when

Subcontractor. An organization, usually a company, working for another organization on some aspect of the project for which the other organization is under contract

Support team. A term used in this book to designate the personnel working on a project who do not report to the project manager administratively

Task. A small part of a project

Task force. An ad hoc group designated to cope with a project, similar to a project organization although frequently staffed with personnel on part-time assignment, usually adopted by a functional organization having only one project or at most a few projects at any given time

Time compression. The act of reducing the planned time for an activity, accomplished perhaps by adding unplanned staff or using overtime

Time and material (T&M) contract. A contractual form in which the customer agrees to pay the contractor for all time and material used on the project, including a fee as a percentage of all project costs

Top down cost estimating. The approach to making a cost estimate or plan in which judgment and experience are used to arrive at an overall amount, usually done by an experienced manager making a subjective comparison of the project to previous, similar projects

Topical review. Any kind of project review devoted to a single topic, such as a final design review or a manufacturing review

Triple Constraint. The term used in this book to describe the three key project objectives that must be simultaneously accomplished—the performance specification, the time schedule, and the monetary budget

Venture organization. The form of organization used in some large organizations where a three- or four-person team, itself functionally organized, is established within the larger organization to develop and commercialize a new product

Work breakdown structure (WBS). A family tree, usually product oriented, that organizes, defines, and graphically displays the hardware, software, services, and other work tasks necessary to accomplish the project objectives

Appendix 3

Planning Checklists for Project Managers

1. Hardware engineering or study
       projects
   system review
   system approval
   system test criteria
   system test plan
   detailed hardware specification
   customer furnished equipment
   power requirement plan
   weight control plan
   breadboard design
   breadboard fabrication
   breadboard test
   block diagram
   schematic diagrams
   circuit diagrams
   conceptual design review
   preliminary design review
   critical design review
   final design review
   prototype design
   prototype fabrication
   prototype test
   design freeze
   drawings freeze
   functional designs

   system logic design
   optical design
   mechanical design
   electronic design
   thermal design
   subsystem hardware
       implementation
   subsystem software
       implementation
   subsystem integration
   subsystem review
   subsystem approval
   subsystem test criteria
   subsystem test plan
   make/buy decisions
   long lead items
   special test equipment
   commercial test equipment
   calibration of test equipment
   software tests
   data reduction plan
   operational software
   subsystem cabling
   system cabling
   installation planning
   experimental development plan

support plans
support instrumentation
facilities
training plans
repair facilities and requirements
inspection
preshipment review
customer inspection
customer acceptance
preparation for shipment
shipment
customer support
qualification test
flight acceptance test
launch support
mission support
personnel recruitment
personnel reassignment
documentation
  project plan
  integrated schedule
  functional requirements
    document
  environmental requirements
    document
  environmental test
    specifications
  environmental test procedures
  environmental test reports
  interface control
  safety plan
  configuration control plan
  failure mode and effect
    analysis
  reliability and quality
    assurance plan
  development test plan
  acceptance test procedure
  calibration plan
  ground data-handling plan
  experiment development plan
  expendables consumption
  engineering drawings and
    drawing list
  parts list

electronic parts acquisition and
  screening plan
materials documentation
manufacturing release
periodic reports (for example,
  monthly)
special reports
final reports
instruction manuals
reviews
  manufacturing review
  management review
  critical design review
  preshipment review
  internal project reviews
  subcontractor progress
    reviews
  customer reviews

2. Programming projects
  applications requirements
  systems requirements
  system inputs
  system outputs
  detailed architectural design
  design specifications
  functional specifications
  security plan
  system test and acceptance
    specifications
  feasibility studies
  file and data requirements
  cost/benefit analysis
  system design
  program design
  system conversion plan
  shipment and delivery
  turnover to operations
  postimplementation reviews
  supplies
  training aids
  review procedures
  hardware requirements
  personnel capabilities
  milestone reviews
  milestone documents

product specifications
project plan
operator instructions
user instructions
library
project index
change control system
data base administration
technical interface manuals
hardware reference manuals
release information
systems reference manual
3. Construction projects
   project work order plan
   personnel assignment
   client review meetings
   public involvement meetings
   environmental report
   meetings with regulatory agencies
   population projections
   residential load projections
   commercial load projections
   industrial load projections
   field checking
   stream gauging records
   stream gauging measurements
   meteorological records
   meteorological measurements
   environmental baseline data
   environmental sampling programs
   ground surveys

photogrammetric surveys
geotechnical explorations
permits
access roads
hydrology
water quality and pollution
   studies
stability analysis
transient analysis
flood control studies
mathematical modeling
easement and permit drawings
   and descriptions
wiring diagrams
piping diagrams
construction cost estimates
estimated operation and
   maintenance costs
life cycle cost analysis
economic studies
value engineering
rate study
notice and instructions to bidders
bid schedules
contract documents
recommendations of awards
shop drawing review
construction surveys
construction engineering
construction observation
construction record drawings

Index